PATTERNS OF SPECULATION
A STUDY IN OBSERVATIONAL ECONOPHYSICS

The collective behavior of economic agents during speculative episodes provides the foundations of this book. Such moments are marked by a special atmosphere of optimism and confidence in the future, which permeates the entire market and society as a whole. Exuberance, ebullience, and bullish are some of the expressions used to describe the up-going phase; whilst words such as uncertainty, fear, contraction and bearish characterize the subsequent down-going phase.

Much as Newton's discovery of universal gravitation could not have derived solely from observing falling apples (despite anecdotal evidence to the contrary), speculative bubbles also require the study of various episodes in order for a comparative perspective to be obtained. The analysis developed in this book follows a few simple but unconventional ideas. Investors are assumed to exhibit the same basic behavior during a speculative episode whether they trade stocks, real estate, or postage stamps. This idea is crucial for setting up a comparative approach. The main objective of this book is to show that behind the bewildering diversity of historical episodes it is possible to find hidden regularities, thus preparing the way for a unified theory of speculation. A theoretical framework is presented in the final chapters, which show how some basic concepts of dynamical system theory, such as the notions of impulse response, reaction times, and frequency analysis, play an instrumental role in describing and predicting various forms of speculative behavior.

Much of the text is written at a level that does not require a background in the technical aspects of economics, finance, or mathematics. It will therefore serve as a useful primer for undergraduate and graduate students of econophysics, and indeed for any reader with an interest in economics as seen from the perspective of physics.

A theoretical physicist by education, Bertrand M. Roehner has been investigating social and economic phenomena during the past 15 years. He is the author of *Theory of Markets* (1995), which explored the space-time structure of commodity prices, and also of *Hidden Collective Factors in Speculative Trading* (2001). The approach used in these books demonstrates how the observational strategy invented by physicists, and successfully applied in astrophysics and biophysics, can be fruitfully applied in the social sciences as well. Professor Roehner has been a visiting scholar at the Harvard Department of Economics (1994 and 1998) and at the Copenhagen Institute of Economics (1996), and he currently serves on the physics faculty of the University of Paris VII.

The great progress in every science came when, in the study of problems which were modest as compared with ultimate gains, methods were developed which could be extended further and further. The free fall is a very trivial physical example, but it was the study of this exceedingly simple fact and its comparison with astronomical material which brought forth mechanics. It seems to us that the same standard of modesty should be applied in economics.

John Von Neumann and Oskar Morgenstern (1953)

Can economics be reconstructed as an empirical science?

Wassily Leontief (1993)

It is a capital mistake to theorise before one has data. Insensibly one begins to twist facts to suit theories, instead of theories to suit facts.

Sir Arthur Conan Doyle (1910)

PATTERNS OF SPECULATION

A Study in Observational Econophysics

BERTRAND M. ROEHNER

CAMBRIDGE
UNIVERSITY PRESS

CAMBRIDGE UNIVERSITY PRESS
Cambridge, New York, Melbourne, Madrid, Cape Town, Singapore, São Paulo

Cambridge University Press
The Edinburgh Building, Cambridge CB2 2RU, UK

Published in the United States of America by Cambridge University Press, New York

www.cambridge.org
Information on this title: www.cambridge.org/9780521802635

First published 2002
This digitally printed first paperback version 2005

A catalogue record for this publication is available from the British Library

ISBN-13 978-0-521-80263-5 hardback
ISBN-10 0-521-80263-6 hardback

ISBN-13 978-0-521-67573-4 paperback
ISBN-10 0-521-67573-1 paperback

Contents

Preface

In the past four years econophysics has become a very popular field among young physics graduates. In discussions I have had with a number of them who visited me I was struck by the fact that for most if not all of them the main objective seemed to be the construction of a (possibly unified) theory. Usually my reaction was to point out that this is precisely what most economists have been trying to do in the past decades and that one of the reasons which may explain their little success was probably the meagerness and inadequacy of the body of evidence on which these theories were erected. In truth, the conviction of these students simply reflected what physics had become in the second half of the twentieth century, namely a highly successful but also strongly structured and more and more theoretically oriented science. It is symptomatic of that trend that some of the most advanced researches concern grand unification and string theories, that is to say two fields which have (so far) little connection with experimental evidence.

Fortunately the "first generation" of econophysics mostly came from statistical physics, a field in which there is a closer link between theoretical and experimental work. On average at previous econophysical conferences at least 80 percent of the models were compared to some kind of statistical evidence. Such a comparison can be made in different ways however. In the early days of thermodynamics a number of basic experiments (such as those by Joule or Boyle) provided firm guide lines and foundations for the establishment of that new science. However, for many advanced modern theories of critical phenomena the theory was developed in an autonomous way and contact with experimental evidence was restricted to a small number of stylized facts. Not surprisingly, a similar methodological divide is also to be found in econophysics.

When a stochastic model of stock price fluctuations is able to reproduce many of the statistical properties of actual price returns this is certainly of importance from the perspective of predictive econometric models, but it gives little information about the possible mechanisms of speculative price peaks. That point can be illustrated by drawing on an analogy with meteorology. A stochastic description of the statistical properties of wind velocities would certainly be almost as difficult to build as one for stock prices, for in both cases there is a mix of short-term fluctuations and large but rare outliers. However, even if we assume that one is able to build a model that can satisfactorily describe the main statistical properties of high-frequency wind data, such a model would give us little understanding about the basic mechanisms that govern wind patterns, such as for instance the Coriolis effect (i.e. the deviation of wind directions to the right in the northern hemisphere). In order to unravel that facet of the phenomenon one would have to compare the pressure field with wind directions at several spatially separated locations. Once observation has established that wind directions are parallel (and not orthogonal) to isobars, one can set out to find a theoretical explanation for that puzzling fact. This, in a nutshell, is the approach that we use in the present book. In short, it should not be considered as a study in theoretical econophysics but rather as an investigation in what may be called "observational" econophysics. In that sense it complements previously published books in econophysics. I prefer the term "observational" to the term "empirical" because it conveys the idea that the objective is not just to collect statistical data but to conduct a series of investigations; in the words of Claude Bernard "the idea must lead the observation."

Needless to say, not all meteorological regularities are as simple and accurately verified as the Coriolis effect. Some effects (such as the formation of tornadoes) depend upon special circumstances, or even on the so-called "butterfly effect"; others are complex phenomena that can hardly be unravelled at a stroke. In the present study we will encounter the same difficulties. Very often it is by decomposing a complex phenomenon into simpler components that we are able to bring in some new light; that analytical methodology plays a crucial role in the present study.

In short, what our approach borrows from physics is a strategy regarding the way observational research should be conducted. Transposed to the social sciences that strategy has important implications: (i) all phenomena will be considered in a comparative perspective; (ii) great efforts will be devoted to find meaningful empirical regularities; and (iii) then, and only then, we will in the last chapters propose a theoretical framework.

Seen with the eyes of a physicist stock markets appear as one of the most tricky problems one can imagine. Indeed, stock price fluctuations result from a fairly complex set of interactions between a large number of agents, as illustrated by the following circumstances which make the problem particularly difficult: (i) There are several sorts of agents, e.g. market makers, short-term traders, and long-term investors, such as mutual fund shareholders, and each class is characterized by specific forms of behavior and characteristic times. There are even different sorts of stocks: growth stocks (for which capital gains are expected) and value stocks (for which substantial dividends are expected) which may have completely different behavior patterns. Thus in 1999 American growth stocks outperformed value stocks by 27 percent, while in 2000 value stocks outperformed growth stocks by 30 percent. (ii) The range, characteristic time and other parameters of the interactions between agents are largely unknown. (iii) A substantial number of the statistical data that would be required are deemed confidential and are not made public. (iv) Stock markets are closely connected to other financial markets, such as the bond, exchange-rate, or real estate markets.

In short, from the point of view of statistical physics, stock markets constitute an open, out of equilibrium system, which involves different sorts of particles and interactions. In addition, because the rules of the market change in the course of time, the whole system is structurally time dependent. This inauspicious picture can once again be illustrated by drawing on the parallel with meteorology. The undertaking is similar to the challenge faced by somebody who tries to build a global meteorological model without knowing the basic laws which govern the interaction of air and water masses – e.g. Boyle's law or the Navier–Stokes equations – and who would, therefore, have to derive them solely from meteorological observations.

When it comes to stock market bubbles the perspective for a theory is even less promising. For the sake of illustration one can consider an analogy with a flood in the Rhone valley. Several circumstances contribute to such a flood: how much snow there is in the Alps, how fast it melts, how much rain fell in the Massif Central, etc. All these factors are more or less unrelated and it is their conjunction which provokes major floods. This makes the construction of a causal theory very difficult. Now, replace water by money; more specifically look at the snow in the Alps as representing the quantity of money in pension funds. In this case the melting of the snow represents the regulation enacted in the United States in the 1970s, which made it easier for pension funds and insurance companies to invest in stocks. Using the same parallel, the rain in the Massif Central would represent venture capital raised in order to support a major technological breakthrough like the Internet. The overall result would be an outstanding stock market bubble. Because the previous factors are largely unrelated a causal theory would be very difficult to build. For coming years the prospect of equity markets depends considerably on whether the American administration will permit Social Security funds to be invested in stocks. This move was initiated in the late days of the Clinton administration, but whether or not it will be implemented is a question for the political scientist, not for the economist or econophysicist.

In short, although fascinating from a historical perspective, the question of stock market bubbles is largely an ill-defined topic for the purpose of scientific investigation, at least if one wants to answer the question "why?" (in this book we rather focus on the "how?" problem).

Far from being discouraging a clear assessment of the difficulty of the task has a number of positive implications for the present study: (i) Instead of restricting my investigation of speculative phenomena to stock markets, I will also consider other speculative markets, such as the markets for real estate, commodities, postage stamps, or antiquarian books. In many respects these markets are simpler than stock markets; for instance they are less open, less structurally time dependent, and involve only one or two types of investors. (ii) Instead of trying to develop a global predictive model I will focus on specific regularities which seem to shape speculative phenomena. By concentrating on the question "how?" the present study prepares the way for a more comprehensive theory of speculation. (iii) Because one is confronted to a multifaceted phenomenon, an important preliminary step is to subdivide it into simpler components. In our flood analogy they would, for instance, consist of studying the consequences for the water level in rivers of snow melting. In contrast to major floods, which are rare, the former process occurs every year and one is thus in a much better position to study it statistically. The approach based on the *separate investigation* of the building blocks composing a complex phenomenon will be further discussed in chapter 1.

There is currently a methodological gulf between econophysics and economics, but there is also a deep cultural divide between economics and sociology which goes back to the origins of the two disciplines. The founding fathers of sociology, such as Emile Durkheim (1858–1917) or Vilfredo Pareto (1848–1923), defined the objective of their field as the *comparative* study of *collective* social phenomena. No model of the individual man was presupposed. On the contrary, information about individual behavior was to be derived from the observation of (large-scale) social phenomena. In more recent times that line of research was continued by sociologists such as Karl Deutsch (1912–), Stanley Lieberson (1933–), or Charles Tilly (1929–). In contrast, by adopting as the cornerstone of the discipline the concept of an individual, rational economic agent, economics started with opposite premises. This had the following far-reaching consequences: (i) The primacy of model building over comparative empirical research. Indeed, since the economic agent was supposed "rational" any economic phenomenon could in principle be derived from a set of axioms. As an illustration it can be mentioned that it is almost impossible to publish an empirical paper in an economic journal (for financial journals the situation is less clear-cut). (ii) A lack of interest in collective phenomena which cannot yet be explained in terms of individual behavior. (iii) A tendency either to ignore or to belittle the connection between economic agents and the society in which they live and work.

By stressing the importance of empirical research, by ignoring the academic borderlines between sociology and economics, econophysicists have initiated a transformation which has been recurrently advocated by renowned economists, such as Clive Granger, Wassily Leontief, or Anna Schwartz. As their recommendations largely fell on deaf ears it became obvious that economics could hardly be reformed from within.

The comparative approach advocated in this book has been used repeatedly in the past, but most often in a fairly unsystematic way. Today, however, thanks to the Internet and to the widespread use of English by statistical agencies all around the world, it has become far easier to collect statistical data from various countries. For instance, it is easy nowadays to download data ranging from real estate prices in Singapore, to office vacancy rates in Houston or assets of equity mutual funds. This represents an historic opportunity for the development of the comparative way in the social sciences. One can hope that in a couple of decades most (non-copyrighted) books will be digitized and made available on line, which would give access to many pre-Internet data as well. In short, the Internet represents a revolution in the social sciences of the same magnitude as observation based on space telescopes in astronomy. In both cases the field of observation is enlarged to unprecedented proportions.

This book summarizes a journey that led me from neutrino physics to the spatial analysis of grain prices, the investigation of Zipf's law, and eventually the study of speculative phenomena. Even if the former topics are only occasionally mentioned in the present book they have necessarily left their mark on my present approach. Even more pivotal was the realization that in economics the main obstacle was not so much the inadequacy of the theories as the lack of definite empirical patterns and regularities. This idea can be illustrated by a personal note. Sometimes colleagues in my department ask my opinion about the topics which should be included into an econophysics curriculum. I guess they expect me to mention the theory of stochastic processes or other subjects in financial mathematics. However, partly because it is my conviction and partly because I am curious to observe my colleagues' reactions, I usually answer: "Well, in my opinion there would be no better preparation to econophysics than to take a course in experimental physics" (as it happens there is indeed in our department an excellent course of that kind). The awareness that observations and facts have to be reshuffled and reorganized before they can be modeled mathematically has guided me throughout the present study.

"Patterns of speculation" could seem to be a vague, catch-all title, but it has the advantage of emphasizing fairly well both the ambition and the limits of my undertaking. The ambition is reflected in the fact that the title does not make any reference to a specific market (e.g. stocks, options, and so on); as a matter of fact my scope includes all markets for which speculative effects can be documented. In short this book argues that there are sufficient similarities between various manifestations of speculation to warrant an all-embracing study. Needless to say, the ultimate objective is to pave the way for a unified theory of speculation.

The word "pattern," on the other hand, emphasizes the limits of the present work. It does not propose a theory of speculation, but if it can convince the reader that such a theory should be possible it will have met its objective. For this purpose we show that behind the bewildering diversity of speculative phenomena there are common regularities. This book describes a number of them, and, even though some are merely qualitative, they are nevertheless of interest in so far as they show that speculative phenomena were not fundamentally different in the nineteenth, twentieth, or twenty-first century. This observation is of fundamental importance, for, if one must propose a new theory every 50 years, economic analysis will be as endless (and hopeless) as the task of Sisyphus.

There are four main parts to this book which, although related, are fairly distinct. The first part presents the main ideas (as seen by the present author) on which are based the econophysical approach to economic phenomena. In particular I emphasize that this approach continues a century-long tradition in empirical and comparative analysis. Biophysics, astrophysics, neurophysics constitute three examples where the approach used by physicists was applied to other fields; as these cases have a much longer historical record than econophysics (biophysics, for instance began in the nineteenth century) it can be of interest to examine these cases more closely in order to identify the factors which brought about their fruitful development. The second and third parts aim to discover qualitative and quantitative regularities in the organization and evolution of speculative markets. The analysis emphasizes that all speculative bubbles are "rational," in the sense that the expectations of investors are consistent with the preconceptions and social climate that prevailed at that time. In other words, the rigid concept of rationality symbolized by the so-called homo economicus will be replaced by an extended concept of *social rationality*. If one looks at speculative bubbles by focusing on social behavior, deep similarities become visible beyond apparently distinct phenomena. In our analysis of qualitative regularities one of our main threads will be the theme of social productivity: many of the changes in the organization of markets came about as an attempt to perform the same function at a smaller social cost. The final part of the book proposes some elements for a theory of speculative price peaks.

The writing of this monograph was an exhilarating journey in the course of which I came to explore many different facets of speculative trading ranging from diamond or postage stamp markets to stock or bond markets. Maybe in some places I have erred; that is almost inevitable if one considers the diversity of the data that needed to be processed. Needless to say, I welcome notification of possible errors or omissions.

Another observation is in order regarding repetitions. Only a few readers will probably read this book throughout, from first to last chapter, and, accordingly, some useful definitions or arguments have been purposely repeated in different chapters.

In an article published in *Europhysics News*, Peter Richmond (2000) wrote that econophysics "offers both the excitement and bewilderment that must have been felt by the pioneers of thermodynamics or quantum mechanics in previous centuries." This view certainly captures an essential aspect of econophysics, namely the enthusiasm of a small and closely knit community of researchers. From its very beginnings econophysics was, in R. Putnam's terms (2000), endowed with a high level of social capital.

For me this community was of vital importance. Thanks to my contacts with other econophysicists, what would otherwise have been a long and lonely voyage far away from home has become an exciting trip; to all I want to express my sincere gratitude. Together Rosario Mantegna, Luis Amaral, and Gene Stanley were the moving spirit behind many innovations and initiatives. Yi-Cheng Zhang and Sergei Maslov were liberal idea providers. Didier Sornette shared with me his enthusiasm and together with Thomas Lux worked at bridging the gap between econophysics and economics. Many thanks to Dietrich Stauffer for his perceptiveness, insight, and invigorating sense of humor. Together with his colleagues at Science et Finance, Jean-Philippe Bouchaud introduced many sophisticated mathematical models. Doyne Farmer provided an essential link between European econophysical centers and the Santa Fe Institute; moreover I much appreciated his lucid approach regarding simplicity versus complexity in economic systems. The views expressed in this book have also benefited from discussions with several colleagues to whom I express my gratitude; let me mention in particular James Feigenbaum, Peter Freund, Taisei Kaizoji, Vasiliki Plerou, Sorin Solomon, and Gilles Zumbach.

I am deeply indebted to several distinguished economists for their unfailing support and interest. Guy Laroque and Edmond Malinvaud have been a permanent source of stimulation since the early 1990s. The encouragements that over the years came from Milton Friedman, Clive Granger, Anna Schwartz, Richard Sylla, and Jeffrey Williamson were highly appreciated. The work of other renowned "comparativists," such as James Foreman-Peck, Gunnar Persson, and Graeme Snooks, convinced me that I might be on the right track after all.

I also express my gratitude to my colleagues at my institute for their daily help and good humor, and especially to Laurent Baulieu, Bernard Diu, Jean Letessier, Annie Richard, and Ahmed Tounsi.

This book is dedicated to Brigitte and Sylvain, my wife and son, whose cheerful encouragement and stimulating support was invaluable.

Bertrand Roehner
Paris, February 19, 2001

Part I

Econophysics

1

Why econophysics?

Johann Gregor Mendel was born in Hyncice, in what is now the Czech Republic, on July 22, 1822. After studying science at Vienna (1851–1853), he became Abbot at Brno (1868). The question of how traits were passed from one generation to the next was at that time extensively investigated by several scientists, but with fairly inconclusive results. Unlike others Mendel studied only one trait at a time and he studied several generations instead of just two or three. He also managed to set apart accidental factors, such as the influence of foreign pollen. It is estimated that in the course of his investigation he observed about 28,000 peas, a figure which attests to the thoroughness of his investigation. He also devoted much time to meteorological observations; in addition to his two celebrated papers on hybridization he wrote nine articles on meteorological questions. This part of his activity is less well known because it did not lead to path-breaking discoveries, but it is interesting to observe that it continued a well-established tradition. Before him several other great scientists, such as Kepler (1571–1630), Descartes (1596–1650), or Lavoisier (1743–1794), had devoted a substantial part of their scientific activity to meteorological studies, without, however, being able to make significant inroads. This short account of Mendel's accomplishments encapsulates several of the themes that we develop in this chapter, such as the emphasis on thorough and systematic experimental work or the classification of scientific problems according to their degree of complexity.

As one knows the term econophysics designates the investigation of economic problems by physicists. It became a recognized field in physics around the mid 1990s when some physical journals began to publish economics studies. The word "econophysics" is a neologism which was coined in 1997 by Eugene Stanley on the pattern of "neurophysics" or "biophysics." However, in contrast to biophysics, which has a fairly clear justification as the study of the physical phenomena (such as for instance osmosis) which play a role in biology, the rationale for a marriage between physics and economics is less obvious. It is the role of this chapter to clear up that point. But, before coming to that, we must discuss two alleged (but nevertheless often mentioned) justifications.

The first one is that some theories developed to deal with complex systems in physics can possibly be applied to economic systems as well. This idea is not new; after all the mathematical framework of the theory of general equilibrium developed by Walras and his followers was largely borrowed from classical mechanics. However, in a general way the idea that a theory can be developed independently of observation seems weird and in any case is completely at variance with physical thinking. The theory of general equilibrium is no exception; it has led to an elaborate formalism which has very few points of contact with observation.

The second justification is the claim that because of their mathematical ability theoretical physicists are in a good position to build economic models. This may have been true before the 1950s when economic teaching was still essentially qualitative and non-mathematical. At that time, the only way to obtain a good mathematical foundation was to graduate as an engineer or a physicist. Several great economists, such as V. Pareto (1848–1925) or M. Allais (Nobel laureate in 1988), were indeed trained as scientists. However, in the second half of the twentieth century, there was an explosion in the number of journals and papers in mathematical economics (see in this respect Roehner 1997, 10–11). One needs only to leaf through a journal such as *Econometrica* to realize that economists are hardly in want of mathematical sophistication.

In this chapter we take a completely different position and argue that what hinders the development of economics is not the inadequacy of the theoretical framework but rather the difficulty of conducting satisfactory observations. To begin with, we trace the elements which in physics permitted a fruitful interaction between theory and observation. In the first two sections we take as our starting point Newton's apple paradigm.

1 Newton's apple paradigm revisited

"[After dining with Newton in Kensington on April 15, 1726] we went into the garden and drank tea under shade of some apple-trees. Amidst other discourses he told me he was just in the same situation as when formerly the notion of gravitation came into his mind. It was occasioned by the fall of an apple as he sat in contemplative mood. Why should that apple always descend perpendicularly to the ground thought he to himself. Why should it not go sideways or upwards but constantly to the earth centre?" This is how William Stukeley (1752) recounts the celebrated anecdote about Newton's apple.

1.1 Newton's apple

Nowadays we are so accustomed to associating the fall of an apple with the concept of gravitation that it is easy to overlook many important aspects of the question. For instance, the very fact that Newton concentrated his attention on the trajectory of the apple is non-trivial. As a matter of fact, if we look at that phenomenon with the eyes of someone such as Descartes or Galileo, who both lived before Newton, we can observe three phases (fig. 1.1):

- The apple breaks loose from the branch possibly because of a sudden gust of wind.
- The apple falls.
- The apple hits the ground.

Of these three phases it is the second which captured Newton's attention, but it is in fact the least spectacular. The fall occurs without a sound and has no incidence on the apple. However during the two other phases, the apple undergoes a visible

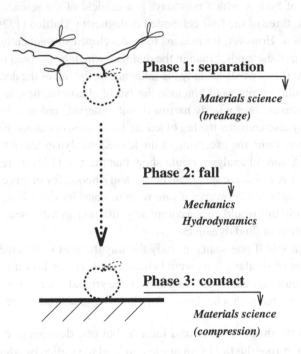

Phase 1: separation

Materials science (breakage)

Phase 2: fall

Mechanics Hydrodynamics

Phase 3: contact

Materials science (compression)

Fig. 1.1. Fall of an apple
Notes: There are (at least) three phases in the fall of an apple, each of which pertains to a different branch of physics. Historically, it was the investigation of phase 2 which proved of paramount importance by leading Newton to his ground-breaking theory of gravitation. In terms of complexity phase 2 is a two-body problem while the two other phases correspond to *N*-body problems. Focusing on that phase constituted an important step in Newton's discovery.

transformation. For instance, the rupture of the link between the stem (the peduncle in botanical terms) and the branch is by no means a trivial phenomenon. Furthermore the fact that the apple may be bruised when it hits the ground is of interest to the gardener for it is well known that a bruised apple cannot be stored for a long time.

Unlike history, science is not concerned with single events; single events can only be described. In order to discover regularities one must consider a *collection of similar events*. Of course it is the word *similar* which is both crucial and difficult to define. For the purpose of illustration let us consider the collection of events which can be considered in relation to each of the phases of the fall of an apple. Needless to say, these clusters of events are not defined univocally; they completely depend upon the phenomenon that one wants to study.

1 *Separation from the tree* If one wants to study the influence of the diameter or length of the peduncle one may consider a cluster of events, comprising the fall of cherries, plums, hazelnuts, and so on. This kind of study would pave the way for the science of breaking away of bodies, which nowadays is a subfield of the science of materials. As a matter of fact, three of the four celebrated dialogues by Galileo (1730) are concerned with that question. However, for reasons to be developed subsequently, Galileo was far less successful in these studies than in the dialogue which he devoted to mechanics.

2 *Fall of the apple* If one wants to study the influence on the fall of the density of the apple, a cluster of similar events would include the fall of chestnuts, figs, lemons, and so on. However these cases cover a fairly narrow density interval, and in order to broaden that interval one may also consider the fall of leaves, hailstones, or cannon balls. As we know, for all these items (with the exception of the leaves) the dynamics of the fall is more or less the same. A careful analysis would show that the $z = (1/2)gt^2$ regularity is better respected for items which are spherical, dense, and smooth, for in this case air resistance can be almost neglected. However if one were to consider the fall of various items in water, wine, or oil, this would open a completely different area of research and eventually lead to the creation of fluid dynamics.

3 *Landing of the apple* If one wants to study the way the fruit is affected when it hits the ground, a cluster of similar events would consist in observing how nuts, oranges, pears, peaches, or tomatoes are damaged when they hit the ground. Like the first phase, this phenomenon is connected with a field nowadays known as materials science.

In short, depending on the phases and factors that one decides to consider, a fairly simple observation like the fall of an apple can lead to studies in what we now know to be different fields of physics. Before we apply the apple paradigm to economic systems, let us come back to Newton. As one knows, his genial intuition was to include the Moon into a cluster of events similar to those of the fall of an apple. This was a brilliant generalization for, at first sight, the Moon seems to be of a quite different nature to that of an apple. A further generalization was made two centuries later by Einstein, when he included a beam of light into the cluster of falling bodies. The expedition led by Eddington during a total eclipse on Principle Island (West Africa) confirmed that, as predicted by the theory of general relativity, the light coming from stars just beyond the eclipsed solar disk was attracted by the gravitational field of the Sun.

1.2 An economic parallel

To many readers our discussion regarding Newton's apple may perhaps have sounded fairly trivial. However, by considering an economic analog, it will soon be discovered that the questions that it raised are at the heart of the problem. Suppose that on a particular Thursday, late in the afternoon, the American Department of Commerce releases inflation figures which turn out to be higher than were expected by the market, say 3.5 percent (in annual rate) instead of 1.5 percent. The next morning the Dow Jones industrials lose 5 percent right at the beginning of the session; fortunately in the afternoon the index regains 3 percent, thus limiting the daily fall to 2 percent. This is a fairly simple event but, as in the case of Newton's apple, one can distinguish different phases.

1 There is the reception of the information. The bad inflation figure comes to the attention of investors through various information means (Reuters headlines, internet, and so on), but it is very likely that interpersonal communication between analysts at various banks and financial institutions plays a critical role in the way the information is eventually interpreted. One can, for instance, imagine that the decision to sell taken by a few stock wizards induces other investors to follow suit.

2 On Friday morning before the opening hour the specialists who are in charge of the 30 stocks composing the DJI discover the level of selling orders that have been sent in during the night. In order to balance sales and purchases they have to set prices which are 5 percent below the previous day's closing prices.

3 Investors who have bought options (either to sell or to buy) over previous days (or weeks) were caught off guard by the unexpected inflation figure. The expiration day of many options is usually Friday at the end of the session, and, in order to improve their position on the option market, the option holders bought heavily in the last hour of the session.

More detailed explanations concerning the organization of stock markets will be given in subsequent chapters, but at this stage it is enough to realize that a certain trigger factor (the announcement of the inflation figure) produces a given result (the fall of the DJI) through a chain of mechanisms each of which represents a fairly complex phenomenon in itself. Phase 1 concerns the diffusion of news amongst a group of people. A cluster of similar events could include the diffusion of other unanticipated news, such as the assassination of Martin Luther King (1968) or the invasion of Kuwait by Iraq (1990), although these are more sociological problems than economic. Phase 2 concerns the procedure of price fixing: to establish, given a set of buying and selling orders, the "best" equilibrium price. A cluster of similar events would include price fixing episodes in other markets, such as commodity markets. Phase 3 concerns the interaction between option and stock markets. A cluster of similar events would include other episodes where stocks and options prices tend to move in opposite directions.

Earlier we saw that the various phases could be linked to different branches of physics. Do we have a similar situation above? As already observed the question raised by phase 1 appertains to sociology rather than economics, and it seems so far to have received very limited attention from economists. The issue of price fixing is a central problem in economics; in the present case, however, one is not in a favorable position because the data about the amount of selling or buying orders would not have been made public. The question raised by phase 3 is perhaps the only one for which satisfactory empirical evidence would be available; nevertheless, to our best knowledge such an investigation has not yet been performed in a systematic way.

By comparison with the case of the fall of an apple, we are in a less favorable situation. Whereas each phase could be linked to a well-defined branch of physics, here there is no organized body of knowledge that one can draw upon. Any investigations which have been devoted to the above issues are not of much help because: (i) they are scattered in various journals and therefore difficult to locate; (ii) they are cumbersome to use because, usually, they do not deal with a single and well-defined issue, but rather with compound and multifaceted problems. Whereas in physics complex phenomena are routinely decomposed into simpler components, each of which is well documented in various physical handbooks, in economics decomposing a multifaceted phenomenon into simpler components (fig. 1.2) is far from being a common approach. In the next section we discuss some historical factors which can explain why physics and economics developed in different ways.

2 Simple phenomena first

"Why, when left in the sun, does ice not soften like butter or wax? Why does the volume of water increase when it is changed into ice? Why is it possible by using salt to make water freeze in summer time?" These are some of the questions raised by R. Descartes in 1637 in his *Discourse on Method* (1824, my translation). These are difficult problems, and even nowadays physics is unable to propose simple explanations for these phenomena. In the methodological part of this work Descartes recommended to decompose every problem into as many parts as were required to solve it. However, this excellent rule was of little help because, as will be explained, the problems that he considered are intrinsically complicated. As a result one cannot be surprised by the fact that the answers provided by Descartes were highly

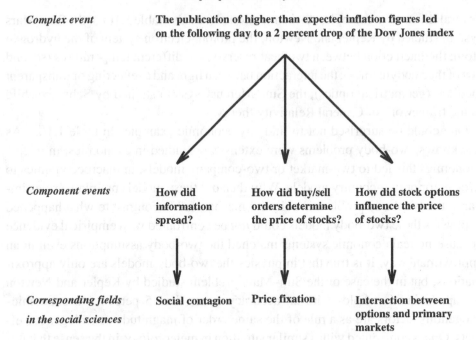

Complex event   **The publication of higher than expected inflation figures led on the following day to a 2 percent drop of the Dow Jones index**

Component events

| **How did information spread?** | **How did buy/sell orders determine the price of stocks?** | **How did stock options influence the price of stocks?** |

Corresponding fields in the social sciences

Social contagion   Price fixation   Interaction between options and primary markets

Fig. 1.2. Fall of the Dow Jones index
Notes: At least three different phenomena can be distinguished in the events that brought about a fall in the Dow Jones index. But in contrast to the fall of an apple the specific branches of economics that would treat these different phenomena are not well developed yet.

unsatisfactory. The only phenomena for which he provided appropriate models were those concerned with the study of refraction and the explanation of the rainbow; we will soon understand why. Half a century later, Newton undoubtedly was more successful and one of the main reasons for that is the fact that by concentrating his investigations on optics and mechanics he set himself more reachable goals. The fall of an apple for instance is what physicists call a two-body problem for it involves only two interacting items, namely the apple and the earth. In this section we understand this notion in a somewhat extended sense. For instance, when a beam of light goes through a prism we will still say that it is a two-body phenomenon because it involves only two items, namely the light and the prism. Somehow in the same spirit, Mendel's experiment with peas can be called a two-body phenomenon because it involved only two types (smooth versus wrinkled) of peas.

2.1 Two-body problems

A little reflection shows that most of the problems that physics and biology were able to solve in the nineteenth and early twentieth centuries were of the two-body type.

Several illustrative examples can be mentioned (see also table 1.1(a)): the Sun–Mars system studied by Kepler and Newton; the proton–electron system of the hydrogen atom; the interaction between two heat reservoirs at different temperatures (second law of thermodynamics); the interaction between light and a reflecting or transparent medium (geometrical optics); the Sun–Mercury system studied by Schwarzschild in the framework of General Relativity theory.

One could be surprised not to find any economic examples in table 1.1(a). As one knows, two-body problems were extensively studied in economics: in micro-economics this led to two-market or two-company models, in macroeconomics to two-sector or two-country models. Why then did these models not play a role similar to the two-body problem in celestial mechanics? In contrast to what happened in physics these two-body models could *not* be confronted with empirical evidence because no real economic systems matched the two-body assumptions even in an approximate way. It is true that in physics the two-body models are only approximations, but in the case of the Sun–Mars problem studied by Kepler and Newton the approximation holds with an accuracy better than 5 percent. In economics exogenous factors are as a rule of the same order of magnitude as endogenous effects. One is confronted with a similar situation in meteorology, in the sense that it is almost impossible to isolate two-body systems which can be studied independently of their environment.

Perhaps some simple economies that existed in the mid nineteenth century in parts of Africa and in some Pacific Islands could have provided closer two-body system approximations. However, unfortunately, to the best of my knowledge, no comprehensive statistical data are available for such societies. In the same line of thought, ant colonies constitute examples of fairly simple economic systems and in these cases it would be possible to generate reliable data by monitoring ant colonies in the laboratory. For instance, colonies of harvesting or gardening ants (Weber 1972) could provide a model of a two-sector economy. Of course it would be a non-monetary economy for which one would have to reason in terms of working time and material output. Such an avenue of research will probably be explored in the future, but at present no data of this sort seem to be available.

2.2 Complexity classification

Table 1.1(b) lists problems in physics, biology, or the social sciences by order of increasing complexity. The Ising model appears at the very beginning of the list of N-body problems for two obvious reasons: (i) it provides a fairly good description of the phenomenon of ferromagnetism and (ii) the analytical solution obtained by L. Onsager (1944) for a two-dimensional Ising model provided for the first time a spectacular illustration of the fact that a short-range interaction restricted to nearest neighbors could generate unexpected effects, such as spontaneous magnetization – below a critical temperature the spins align themselves and collectively generate a non-zero magnetic moment. From the perspective of economics this example is of great interest because it proves that a simple interaction between agents can bring about non-trivial collective phenomena. Table 1.1(b) shows that the basic problem of macroeconomics, namely the interaction between N sectors of an economy is mathematically of a level of difficulty which is far above that of the two-dimensional Ising model. Indeed, whereas the Ising model involves only one sort of interaction, an N-sector economy involves a variety of interactions. What makes things worse is the fact that these interactions are usually time dependent. For equity markets there are also different kinds of interaction: banks, pension funds, insurance companies, or equity option speculators do not have the same goals nor the same reaction times.

Table 1.1(b) can help us to better understand some of the observations that we made earlier in this chapter: (i) Let us first consider the questions raised by Descartes that we mentioned at the beginning of this section. The question of water versus ice volume is of complexity $C2$; the question of butter versus ice is at least of level $C2$ and probably of level $C4$ if butter contains several kinds of molecules which are relevant. One understands why Descartes was more successful in optics which is of level $C1$. (ii) The gravitational interaction between the earth and a falling body belongs to class $C1$, whereas materials science belongs to class $C2$; thus one can understand why Galileo was able to unravel some of the laws that govern the movement of a falling body but was less successful in his studies regarding the coherence of solids. (iii) As we pointed out at the beginning of this chapter, many brilliant scientists such as Kepler, Lavoisier, or Mendel made extensive studies in meteorology without, however, being able to offer any decisive and ground-breaking contribution; this becomes understandable on account of the fact that meteorology is of complexity level $C4$. A complementary complexity criterion is given in table 1.2.

2.3 The role of time: Simon's bowl metaphor

Among the economists of the 1960s and 1970s, the work of H. Simon (Nobel prize laureate in 1978) stands somewhat apart. His approach to economics was largely influenced by system theory, and it is therefore not surprising that he gave much attention to the question of complexity (Simon 1959, 1962). He proposed the following example in order to illustrate the fact that dynamical problems are inherently far more difficult to study than systems in equilibrium. This is a distinction that we have not made so far and which complements the results in tables 1.1(a),(b). Simon's (1959) argument goes as follows: Suppose we pour some viscous liquid (molasses) into a bowl of very irregular shape, what would we need in order to make a theory

Table 1.1(a). *Two-body problems in physics and biology*

Interaction between:	Century	Scientists
Earth–falling body	16th	Galileo (1564–1642)
Light–glass	17th	Snell (1591–1626), Descartes (1596–1650), Newton (1642–1727)
Sun–Mars	17th	Kepler (1571–1630), Newton
Static gas–container	17th	Boyle (1627–1691)
Flowing fluid–pipe	18th	Bernoulli (1700–1782)
Two heat reservoirs	19th	Carnot (1796–1832)
Liver–pancreas	19th	Bernard (1813–1878)
Two alleles of a gene	19th	Mendel (1822–1884)
Proton–electron	20th	Bohr (1885–1962)
Sun–Mercury	20th	Schwarzschild (1873–1916)

Notes: The scientists' names are given in order to specify to which phenomenon and law we refer.

of the form the molasses would take in the bowl? If the bowl is held motionless and if we want only to predict behavior in equilibrium, the single essential assumption would be that under the force of gravity the molasses would minimize the height of its center of gravity. However, if we want to know the behavior before equilibrium is reached, prediction would require much more information about the properties of molasses, such as for instance their viscosity, density, and so on.

How should Simon's metaphor be understood from the perspective of the classification in table 1.1(b)? Molasses, like Descartes' butter, would belong to level $C2$ if there is only one preponderant interaction or $C4$ if there are several competing interactions. Yet, as pointed out by Simon, the equilibrium problem in a gravitational field is rather a two-body problem. The reason is easy to understand; in equilibrium the interaction between the molecules does not play any role, which means that the bowl of molasses is not fundamentally different from Newton's apple. More specifically the general laws for a fluid in equilibrium also apply to molasses. Alternatively, in a dynamic situation the interactions play a key role and the problem is far more complicated.

2.4 Simple aspects of complex systems

Table 1.1(b) seems to convey a pessimistic view regarding the prospects for economic analysis. Is the situation really hopeless? Not at all. A parallel with meteorology can help us to understand why. Like economics meteorology is of complexity level $C4$; as a matter of fact obtaining an accurate model of the interaction of air and water masses on a world-wide basis is an almost impossible task. Arguments

Table 1.1(b). *Problems of increasing complexity in physics, biology
and the social sciences*

Level of complexity	Problem
C1	*Two-body problems (see table 1.1(a))*
C2	*N-identical body problems with interaction between nearest neighbors*
	Interaction between N spins (ferromagnetism Ising model)
	Interaction between nucleons in complex nuclei
	Interaction within a population of bacteria belonging to the same species
	Interaction between N grain markets
C3	*N-identical body problems with a long-range interaction*
	Interaction between N neurons of same type
	Interaction between N similar investors in equity markets
C4	*N-non identical body problems (several interactions)*
	Interaction of air and water masses (meteorology)
	Interaction between N genes (morphogenesis)
	Interaction between different kinds of neurons in the brain
	Interaction between N words (linguistics)
	Interactions in a colony of bees or ants
	Interaction between N sectors of an economy (macroeconomics)
	Interaction between N national economies (international trade)
	Interaction between various kinds of investors in equity markets
	Interaction between states (international relations)

Notes: In the present classification complexity is understood from a mathematical perspective. What gives some confidence in the pertinence of this classification is the fact that it is corroborated by the historical advancement of science: the understanding of systems of low complexity historically preceded the comprehension of systems belonging to higher complexity levels. For complex systems there are richer forms of collective behavior, but, alternatively, it is more difficult to get analytical results which clearly state the conditions under which these forms of collective behavior will occur. It should be noted that, apart from the type of interaction, there is another essential difference between, for instance, a complex nucleus and an equity market; by contrast with the nucleus for which the properties of the interaction are experimentally well known, for equity markets the interactions between various investors are basically unknown.

about the chaotic nature of these phenomena may even suggest that this objective is altogether unreachable. Nevertheless there are many specific meteorological *mechanisms* which can be understood fairly simply, for instance the fact that in the northern hemisphere wind directions are deflected to the right with respect to gradient lines of the pressure field is known as the Coriolis effect. As it happens the Coriolis force is a standard phenomenon in classical mechanics; but, even without any prior knowledge (that is to say leaving aside for a moment what we know from laboratory experiments), it can also be discovered by carefully analyzing wind

Table 1.2. *An objective measure of biological complexity*

Organism	Fifty percent lethal dose of X-rays [rad]
Viruses	100,000
Bacteria	3,000
Mamalian cells	100

Notes: Table 1.1(b) is of little usefulness if one wants to compare the complexity of various biological organisms for all of them would belong to the *C*4 complexity level. The criterion used in this table is based on the assumption that more complex systems are more sensitive to disorders brought about by radiations. The proportion of surviving organisms is a decreasing exponential function of the dose: the 50 percent lethal dose corresponds to the survival of one half of the population.
Source: Encyclopédie Internationale des Sciences et des Techniques (Paris 1973).

directions at spatially separated meteorological stations. The fact that wind directions are deflected to the right at any time and anywhere in the northern hemisphere suggests that this effect must be due to a basic permanent factor (such as the rotation of the earth). Naturally such an analysis will by no means be straightforward because there are many disturbing factors (thermal effects, mountains, and so on) which must be discarded before the main effect can be isolated. In short, even though the construction of a global theory is a very difficult challenge, it is possible to discover regularities which provide at least partial understanding.

We believe that the situation is very much the same in economics: although the construction of a global model of the world economy may be out of reach, it is possible to build fairly simple models of specific economic phenomena. Such models provide partial understanding but do not permit global prediction, and this is why this objective should be set aside at least temporarily. Such a view was similarly expressed by Schumpeter (1933) in the first issue of *Econometrica*: "We should still be without most of the conveniences of modern life if physicists had been as eager for immediate applications as most economists are and always have been."

The following section offers some practical hints for the search of regularities.

3 From plausible reasons to regularities

Our objective is to generalize the rules that guide experimental investigation to cases where no experiments can be performed and one has to rely on collecting observations. Apart from economics, this discussion also concerns other observational sciences, such as astrophysics, geodynamics, or meteorology. The main difference between these sciences and economics is that the former can rely on physical laws that were discovered independently whereas the basic laws that govern the social behavior of economic agents are still unknown. To begin with we consider a very simple example.

3.1 The pot of yoghurt paradigm

Suppose you spend your vacation in a mountain resort at an altitude of 1,000 meters. You take a pot of yoghurt from the fridge and observe that the thin metallic cover that forms the lid has a convex shape (i.e. the center is higher than the rim). Since you have never observed this before you wonder what the explanation may be. A number of possible explanations can be tentatively proposed; one can mention the following for instance:

1 Some kind of fermentation in the yoghurt has led to the production of gas which has increased the pressure inside.
2 The yoghurt was upside down in the truck that brought it from the dairy to the grocery; this gave to the metallic cover a convex shape which it retained.
3 The engine of the refrigerator contains magnetic parts which exerted a force on the metallic cover.
4 When the pot was sealed at the diary two or three days ago the atmospheric pressure was higher. As the pressure fell the air contained in the pot provoked the convex shape.
5 At an altitude of 1,000 meters the atmospheric pressure is lower than in the valley where the dairy is located. As the truck climbed the mountain side, outside pressure decreased and the air contained in the pot expanded provoking the convex shape.

Before we begin to discuss these different explanations an important comment is in order. In previous sections of this chapter we emphasized that a single event cannot be investigated scientifically, it can only be described; only clusters of events can be scientifically explained. As a matter of fact all the previous explanations are plausible. In order to decide which one is correct, additional evidence must be collected. For instance, one could look at the milk bottles. If their lids have the same shape this would exclude the first explanation (unless one assumes that there was a similar fermentation for the milk, an assumption which although unlikely cannot be completely ruled out). If the convex shape can already be observed in the grocery this would reject the third explanation (unless one assumes that the refrigerator in the grocery or in the truck also generates a magnetic field); of course if one can determine that the pot has an aluminum cover the explanation would definitely be rejected. By examining the aspect of the cover it is perhaps possible to determine whether it was upside down in the truck (or at the dairy). In short, it may be assumed that by a number of additional observations one is able to eliminate (at a reasonable level of confidence) all explanations except the last two.

As one knows each of these phenomena may actually play a role and it is not a simple matter to decide which one is preponderant. The average pressure difference between sea level and an altitude of 1,000 meters is about 110 hectopascal (*Quid* 1997, 107). However the pressure variation between an anti-cyclone and a depression is of the order of 50 hectopascal. But this does not necessarily mean that effect number 5 is preponderant, for the dairy is certainly not located at sea level; if it is located in the valley at an altitude of 600 meters the two effects can be of the same magnitude. If the location of the dairy can be known this will of course facilitate the solution of the problem. If however it cannot be known (which is a fairly common situation in a field such as economics where many data remain confidential) one could try to obviate that lack of information by carrying out additional observations. For instance one could repeat the same experiment of buying yoghurt jars over several days in a row; if the date on which the yoghurt was produced is indicated on the pot the contribution of effect number 5 can be derived (at least in a relative way) from daily records of atmospheric pressure. Usually however the production date is not indicated and one must be content with a reasonable guess, for instance one week prior to the "best before" date.

The reader may perhaps think that an observation regarding a pot of yoghurt hardly deserved such a detailed discussion. In the following paragraph we will give some more academic examples. In any case the previous analysis illustrates and emphasizes two points which are of great importance in economic analysis: (i) It is impossible to reach any definite conclusion by considering a single observation; in order to be able to observe the phenomenon under different circumstances one needs to analyze a collection of similar events. Although this rule may sound fairly obvious to any physicist it is often forgotten in economic analysis as attested by the huge number of books which were devoted to the analysis of the crash of 1929 (a single event). (ii) In economics it is usually not possible to control or even to know all factors which may affect an observation. In many cases hard knowledge has to be replaced by an educated guess. As a result the confidence level of the proposed explanation is usually lower than in physics. However, by including more observations into the cluster of events, it is often possible to improve the level of confidence. For instance, by selecting a period of observation during which atmospheric pressure remained almost constant, it is possible to get rid of changes in atmospheric pressure. In short, by progressively extending the cluster of similar events under investigation, it is possible to converge toward a reliable model.[1]

---

[1] For the application of that approach to a comparison of the spatial distribution of prices in China and France, see Roehner and Shiue (2001).

3.2 Plausible causes versus scientific explanations

In scientific explanation one can distinguish three levels. The first level corresponds to enumerating plausible causes; the second consists in determining which of these causes actually played a role. This second level corresponds to what we call the search for regularities. The third level implies the construction of a theory that can account for a cluster of similar events. For the yoghurt example, the first level consisted in listing five plausible factors; the second level was reached when after a careful discussion we were able to prove that only two of these factors indeed played a role. The third level consists of proposing a quantitative theory that would explain not only the inflation of yoghurt jars but would also account for the deformation of orange juice cartons and other containers.

For the purpose of illustration let us consider a more dignified example, namely the much debated question of the disappearance of the dinosaurs. First of all it must be noted that since this is a single event it does not really admit a scientific explanation. One can list a number of plausible or real causes but it will never be possible to build and test a comprehensive theory. Among the plausible causes that are most often mentioned one can cite a change of climate, lack of food, gigantism, or the fall of a meteorite.[2] The second level of explanation would consist in identifying those factors which actually played a role. For instance if one found meteor craters corresponding to impacts that occurred at the beginning of the period when the population of dinosaurs began to decline, this would lend some credibility to that argument; although it does not really prove that there was indeed a link between the two events. For this purpose one would have to build a model which would quantitatively specify the expected changes in climate and vegetation. Needless to say such a model must then be tested on *all* large known meteor impacts; if the changes in the years following the impact are in agreement with the model, then one can proceed with additional confidence to apply it to the case of the dinosaurs.

It is now time to consider an economic example. In January 1979 a one-carat, flawless, gem diamond cost 20,000 dollars on the Antwerp market; in February 1980 the same diamond was worth 60,000 dollars; subsequently that price plummeted to 40,000 dollars in January 1981 and 20,000 dollars in January 1982 (Diamonds 1988, *The Economist*, Special Report No. 1128). What explanations did diamond experts put forward in order to explain such a sudden price peak? In 1988 an authoritative economic assessment of the past decade was made by the president of the Gemological Institute of America, a well-respected institution in the industry (Boyajian 1988). His explanation goes as follows. In 1976 Israel was a young but

---

[2] In the same spirit Lieberson (2000) lists a number of plausible reasons for the decline of dress hats; a very instructive exercise!

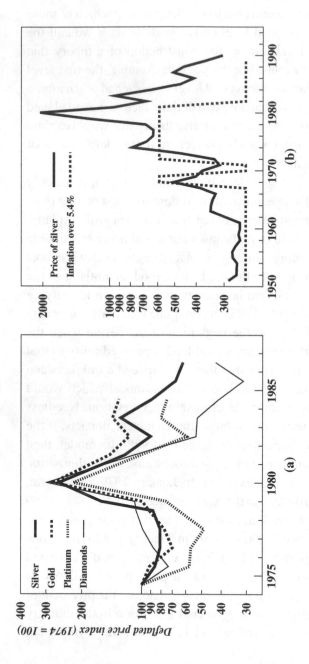

Fig. 1.3a. Price peaks for diamonds, gold, platinum and silver

Notes: Gold, platinum and silver prices are expressed in dollars per ounce troy; for diamonds the price is for a one-carat, G clarity, flawless diamond. The bubble and its collapse occurred almost simultaneously in the four markets. That coincidence contrasts with a rather weak correlation between the four prices in normal times; thus, between 1993 and 1998, the prices of gold and silver were almost uncorrelated. The prices of cobalt and palladium display a similar peak but have been omitted for the sake of clarity.

Sources: *International Financial Statistics* (International Monetary Fund 1979, 1991), *Journal des Finances* (October 26, 1978), *The Economist* (April 5, 1980), Diamonds 1988 *The Economist*, Chalmin (1999).

Fig. 1.3b. Deflated price of silver and inflation rate

Notes: Vertical scale: annual price of silver in New York expressed in 1980 cents per ounce troy; in terms of daily prices the maximum would be at a fairly high level (4,500 cents/ounce on January 8, 1980). The inflation rate refers to the US consumer price index; the dashed line represents a two-valued function which is equal to 1 when the inflation rate is over 5.4 percent and to 0 when it is below. The graph shows that the price peak was largely the result of a flight from inflation and that a 6 percent rate seems to be a critical level in that respect.

Sources: *International Financial Statistics* (International Monetary Fund 1979, 1991); Historical Statistics of the United States (1975).

rapidly growing diamond center. Anxious to promote the diamond trade which represented 40 percent of Israel's non-agricultural exports, the Israeli government supplied several banks with huge amounts of money at very low interest rates to be passed on to diamond manufacturers so they could build their inventories of rough diamonds. But because of the rapid increase in diamond prices many diamond cutters found it more profitable to hold their rough diamonds rather than to cut them. Thus, they created an apparent demand that really did not exist. Soon the speculative fever spread to Antwerp and New York, the two other main diamond centers. At first sight such an account sounds plausible. However, if one looks at it more closely it becomes clear that there are several loopholes in the argument. First, it assumes a rapid increase in diamond prices, which is precisely what one wants to explain; second, one may wonder why the speculative fever spread to Antwerp and New York where no low interest government loans were available. Third, the American and Japanese markets for polished diamonds represented at that time about 60 percent of the world market and therefore one would expect them to play a leading role. Fourth, one would like to know the total amount of the low interest loans made by the Israeli government to the cutters in order to compare it to diamond world sales.

These remarks already make the proposed explanation somewhat doubtful but the decisive point is the fact that the diamond price peak was by no means an isolated phenomenon. As a matter of fact cobalt, gold, platinum, palladium, and silver all experienced a huge price peak, which paralleled almost exactly the one for diamonds (fig. 1.3(a)). Accordingly, one can hardly accept an explanation which concerns only one of these goods; there must have been a factor which explains the other price peaks as well. That factor is the rate of inflation in western countries, which provoked purchases of all kinds of tangibles from precious metals to antiquarian books or postage-stamps (see fig. 1.3(b)). A more detailed analysis is to be found in a subsequent chapter.

In short we see that the scenario proposed by diamond experts was wide of the mark; the injection of capital into the Israeli diamond market cannot be disputed of course, but it had only a negligible impact on the diamond market. Needless to say, similar *ad hoc* explanations were proposed for the other price peaks as well. Thus, the price peak for silver is currently "explained" by the failed attempt of Nelson B. Hunt, a Texan billionaire, to corner the market. Numerous books and articles have developed that thesis; see for instance the article "The silver coup that failed" which appeared in the *Financial Times* (March 29, 1980) or the book by S. Fay (1982).

The previous examples are interesting in another respect. They show that there is a natural tendency to favor short-term, anecdotal evidence at the expense of structural, statistical explanations. Both the Israeli diamond episode and the Hunt silver story occurred in the year preceding the price peak and these anecdotes are of course intuitively more appealing than long-term structural causes, such as the role of inflation. Once again we see that it is only by considering a cluster of similar events that it is possible to get at the roots of the phenomenon.

The reader will perhaps think that we overdid the point by emphasizing the anecdotal character of the explanations put forward by economic experts. How can experts rely on such flimsy evidence? In order to judge for himself (herself) the skeptical reader can have a quick look at the headline news of major economic and financial news agencies, such as Reuters Business News, Reuters Securities, AP Financial, and so on; they can be found on several internet sites. He (she) will perhaps be surprised to see that there is a flood of qualitative statements, but very little global evidence; for instance such important pieces of information as the average price–earnings ratio of the NYSE (New York Stock Exchange) or NASDAQ, the number of bankruptcies in past quarters, the ratio of corporate bond downgrades to upgrades (an assessment of the financial situation of corporations) are hardly ever given.

3.3 Regularities

From a methodological perspective the strong emphasis econophysics has put on the search for regularities is probably one of its most important innovations. In contrast, for most economists a quantitative regularity is considered of no interest unless it can be interpreted in terms of agents' motivation and behavior and has a clearly defined theoretical status. In order to illustrate that difference in perspective let us consider an important result obtained by an econophysical team (Plerou *et al.* 1999). These authors analyzed a huge data base containing annual research and development expenditures (S) for science and engineering in 719 American universities. More specifically they studied the way in which the fluctuations in the growth rate $g(t) = \ln[S(t+1)/S(t)]$ depend on the size of $S(t)$. They have found that the standard deviation $\sigma(S)$ of the fluctuations is proportional to $1/S^{0.25}$. Two circumstances explain why this is a result of outstanding interest: (i) It is fairly robust with respect to the measure of size that is used, whether for instance the expenditure or the number of papers published. (ii) There is a similar law for the growth rate of firms, with an exponent which in this case is 0.17 instead of 0.25. Such a result can be considered as a first step toward a comprehensive theory of growth.

Only a few economists would share that opinion however. In their eyes the previous result has two major shortcomings: (i) it does not fit into any existing theory; (ii) it has no clear interpretation in terms of optimization and rational choice.

3.4 Circumstantial causes versus structural factors

We have already seen that economic experts tend to favor short-term circumstantial causes at the expense of long-term structural factors. Far from being specific to economics this tendency is in fact common to all social sciences, and for an obvious reason: short-term causes seem intuitively far more convincing that structural explanations. Here is a non-economic illustration. During the 1980s and 1990s there have been numerous clashes and even large-scale massacres in Burundi and Rwanda. For each of these episodes one can invoke numerous circumstantial causes, such as an attempted coup, the assassination of a prime minister, and so on. On the structural side one can notice that Burundi and Rwanda have the highest population density in Africa coupled with a high increase rate. In 1995 the density was about 250 people per square kilometer for Burundi and 310 for Rwanda as compared to 110 for Nigeria and 35 in South Africa (*Quid* 1997, 1056). Intuitive reasoning is unable to take into account such structural factors except as vague background cause. In order to assess the role played by these factors one needs a comprehensive quantitative model.

3.5 Models need accurate empirical targets

The main disincentive to improve the handling and use of data is that the [economic] profession withholds recognition to those who devote their energies to measurement. Someone who introduces an innovation in econometrics, by contrast, will win plaudits.

(Anna J. Schwartz (1995))

Nobody would doubt that experimental research is an essential part of physics. Economics, in contrast, does not encourage "experimental" research (quasi-experimental would of course be a more adequate term). The above citation by A. Schwartz describes fairly well this situation. The bias in favor of theoretical work is due partly to a trend that is common to different fields and partly to a tradition which is proper to economics. The first factor is related to the well-known fact that the productivity of a researcher, as measured by his (her) annual output of published papers, is much higher for theoretical work than for experimental research. But for economics this trend was strongly reinforced by its own tradition: in earlier centuries only few data were available and economists developed the habit of replacing genuine observations with "Gedanken" experiments. For instance, instead of observing the actual behavior of real estate investors, economists tried to imagine how, from a rational perspective, they *should* react. Unfortunately there is often a wide gap between expected and actual behavior. The neglect of empirical work was denounced by several great economists, such as M. Friedman,

C. Granger (1991), W. Leontief (1982, 1993), or A. Schwartz, but this did not reverse the trend. From its very beginnings econophysics recognized the importance of empirical research. Many empirical studies (several of them are discussed in subsequent chapters) which would never have found their way into economic journals appeared in physical journals. Over the next few years this should lead to the accumulation of a wealth of regularities and empirical knowledge which may provide the much needed guideposts and landmarks for the building of models and theories.

Why are such guideposts so indispensable? The answer can be illustrated by the following example. By 1999 no less than ten econophysical models were reported in the pioneering book by Mantegna and Stanley (2000) which were able to explain the main stochastic properties (e.g. rapidly decreasing auto-correlation or the fat tail of the distribution function) of stock price changes. To that number an equal number of models developed by economists or financial analysts should certainly be added. When more than 20 models are able to explain the main empirical features it is clear that the constraints are too lax and the target too big. Four centuries ago the situation was the same in astronomy: until Tycho Brahe's observations, the Ptolemy model was as acceptable as the Copernicus model.

4 Conclusion

The message of this chapter comprised three main points: it emphasized the crucial role of observation; it argued that major progress can come from the study of simple problems; and, finally, it insisted on the importance of comparative analysis.

4.1 The primacy of observation

Most of the Nobel prizes in physics have been awarded for experimental discoveries. Moreover, to the best of my knowledge, not a single one has been attributed for a theory which could not or had not been confronted with experimental evidence. The situation is completely different in economics. As a matter of fact several Nobel prizes have been awarded for work which remains completely theoretical. Just to mention three examples, one would in vain search for any statistical test in the works of P. Samuelson (Nobel prize in 1970), G. Debreu (Nobel prize in 1984), or M. Allais (Nobel prize in 1988). This striking contrast emphasizes the fact that observation and experimental evidence have a completely different status in physics and economics. It is true that (quasi-) experimental research in economics is more difficult to conduct than in physics, but this should rather encourage us to devote more time, energy and patience to such research.

4.2 *"Modest goals"*

Throughout its history, confronted as it was with the expectations of governments, traders and investors, economics has had a tendency to take up global problems and to develop predictive models. But such ambitious objectives diverted economists from the patient work of collecting and organizing empirical evidence, which would have permitted them to discover significant regularities. Once again a parallel with meteorology may be enlightening. In meteorology, as in economics, one of the main objectives is to make reliable forecasts; however this did not prevent meteorologists from establishing firm foundations before trying their hands at making forecasts. Thus, in the 1830s, the comparison of wind directions at several places uniformly spaced over a large area constituted one of the major tasks of comparative meteorology (also called synoptic meteorology). Such a study had no direct practical application in terms of forecasting ability, but it permitted the establishment of Coriolis law, which governs the movements of air and water masses at the surface of the earth, and paved the way for further progress. This is the procedure advocated by J. von Neumann and O. Morgenstern in the citation mentioned at the beginning of this book: to set oneself modest goals and to advance step by step from one clearly understood question to the next.

4.3 *Clusters of events and comparative analysis*

Taking Newton's apple paradigm as our starting point we emphasized that even such a simple event as the fall of an apple has in fact many different facets. In order to analyze it in a meaningful way one must first *decompose* it into simpler components and, when this has been done, try to conjecture which components correspond to "simple" phenomena. Once a specific phenomenon has been selected the real empirical investigation begins, which consists of collecting evidence about a cluster of similar events. By comparing the evidence from these events it becomes possible to determine which factors play a crucial role and which ones are merely incidental.

In the previous procedure the critical step of selecting the events to be investigated depends to a large extent on the criteria used to define the degree of complexity (or simplicity) of a system. In this chapter we proposed two main criteria: (i) the number of interactions between the elements composing the system; (ii) the status of the system with respect to time, that is to say whether or not the system can be considered in equilibrium.

The key idea that only a comparative analysis of a cluster of similar events can reveal meaningful regularities was illustrated through several examples. In particular we saw that the "explanations" put forward by experts are often no more than fanciful anecdotes. The approach delineated in this chapter will be applied repeatedly in subsequent chapters. But, before coming to that, we give in the next chapter some brief indications about the historical development of econophysics.

2

The beginnings of econophysics

In the late eighteenth century G. Galvani and A. Volta investigated the contraction of frog muscles subjected to electrical stimulation. This study culminated in Volta's invention of the electrical battery and can be considered as one of the first studies about physical phenomena in living organisms. Subsequently, in the second half of the nineteenth century, the German electrophysiologist Dubois-Reymond was able to measure the minute currents generated in muscles. These studies about the electric properties of muscles opened the way for a field which after 1920 became known as biophysics. One of the peculiarities of biological phenomena is the fact that there are many more parameters than in physics and that it is more difficult to control them adequately. As a result it is rather difficult to stage experiments that are truly reproducible. In order to be successful early biophysicists had to possess the ability to separate complex biological problems into segments amenable to physical interpretation and experimental testing. As demonstrated by fig. 2.1 the first books and journals whose titles contained the word "biophysics" appeared in the decade 1921–1929. However the real emergence of biophysics as a major cornerstone of modern biology may be attributed to the spectacular success of biophysical tools in unraveling the molecular structure of the deoxyribonucleic acid (DNA) and in establishing the detailed structure of proteins such as hemoglobin.

What is our purpose in recalling the history of biophysics? It seems to us that the analogy between biophysics and econophysics is much deeper than the mere similarity in their denomination. As does biology, economics deals with multifaceted phenomena and as a result any investigation must be preceded by the crucial step of separating them into segments, modules, and components which can be studied individually. Our brief account of the history of biophysics also shows that the emergence of a new scientific discipline is a long process which should be measured in decades rather than in years. The formal emergence of biophysics as a new field in the 1920s was preceded by a long period during which the field was developed

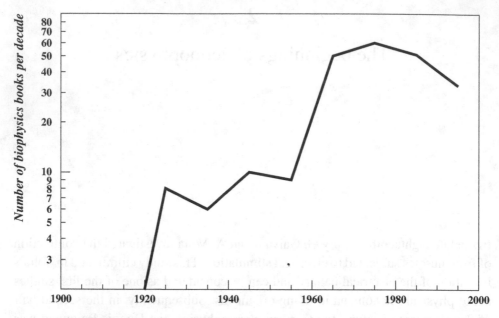

Fig. 2.1. Number of books published per decade whose title contains the word "biophysics"
Notes: "Introduction to biophysics" was the first book in this list; for books in French and German the first dates were 1912 and 1922 respectively. There were two periods of rapid development namely the 1920s and the 1960s. The data are based on the on-line catalog of the Harvard Library.

by individuals with a personal inclination for cross-disciplinary research. The same process can be observed in econophysics; as will be shown in more detail in the next section several great economists of the late nineteenth or early twentieth centuries were educated as physicists.

1 Pre-econophysics

By browsing through the work of great economists like Ricardo, Marx, or Keynes we are able to get a feeling for what economics was like at their time; in particular it is striking that there are almost no statistical data and very few mathematical formulas. As a matter of fact, before World War II economics was mainly a qualitative discipline. At that time the only way to get a training in mathematics and statistics was to be educated as a scientist or an engineer. Therefore it is hardly surprising that the few economists who made use of statistical and mathematical tools were former scientists. This assertion will be illustrated by the examples of A. Quételet (1796–1874), L. Walras (1834–1910), V. Pareto (1848–1923), and M. Allais (1911–).

1.1 Pre-econophysicists

1.1.1 Quételet (1796–1874)

Adolphe Quételet was born in 1796 in Ghent (Flanders). He received his doctorate for a dissertation on the theory of conic sections (1819). In 1823 he came to Paris to study astronomy with Arago and Laplace. After returning to Brussels he established methods for the comparison of physical as well as social data. In 1835 he published a book entitled: *Sur l'homme et le développement de ses facultés. Essai d'une physique sociale* [A treatise on man and the development of his faculties. A study in social physics]. Quételet also published several astronomical studies (1841, 1870). Throughout his life he was an astronomer with a strong personal interest in the social sciences and particularly in demographic questions. Back in the nineteenth century, scientific specialization was less pronounced than it is nowadays, which made it easier to cross the boundaries between different disciplines.

1.1.2 Walras (1834–1910)

Léon Walras was a French economist whose work *Elements of pure economics* was the first comprehensive mathematical analysis of economic equilibrium. After failing the entrance examination to the Ecole Polytechnique in Paris he entered the Ecole des Mines, which, like the Ecole Polytechnique, was a highly regarded institution for the education of engineers. In 1870 he was appointed to the chair of political economy at Lausanne (Switzerland); he published his main works between 1874 and 1877.

1.1.3 Pareto (1848–1923)

Vilfredo Pareto was an Italian economist and sociologist. He was one of the first economists to use mathematical models together with statistical evidence. In particular he discovered the so-called Pareto law which describes the distribution of revenue and wealth in a population. Not surprisingly, he was educated as an engineer at the Instituto Politecnico of Turin, where he graduated at the top of his class. For some years after graduation he worked as a civil engineer for the Italian Railway Company. In 1893, 23 years after Walras, he was appointed professor of political economy at the University of Lausanne. Pareto turned to sociology somewhat late in his life; his monumental *Traité de sociologie générale* [A Treatise on General Sociology] appeared in 1917 (he was then 69). This title is somewhat misleading however, for it is by no means a general panorama of sociology but rather a very personal and original investigation of social customs and superstitions. The real topic of the work is better reflected in the title of the English translation which appeared in 1935 namely, *The Mind and Society*.

1.1.4 Allais (1911–)

Maurice Allais is a French economist who was awarded the Nobel prize in 1988. He was a pre-econophysicist to an even greater degree than Walras and Pareto; not only was he educated as a scientist but throughout a large part of his active life he undertook research in physics. Unlike Walras he was admitted to the Ecole Polytechnique and graduated at the top of his class. After that, this time like Walras, he spent two years at the Ecole des Mines. He wrote a large part of his work in economics in the 1940s. Then in 1953 he started a campaign of experiments involving a special kind of Foucault pendulum; between 1957 and 1959 he published ten papers on the oscillations of that pendulum. This research was interrupted in 1959 when funding was discontinued. Allais then returned to his research in economics, but he did not forget his interest in physics altogether, and in 1997 he published an account of his earlier physical investigations.

1.1.5 Other pre-econophysicists

Many other examples of pre-econophysicists could be mentioned in addition to those above: (i) the American astronomer Simon Newcomb (1835–1909) wrote a book entitled, *Principles of Political Economy*; (ii) a pioneering dissertation by Louis Bachelier (1870–1946) appeared in 1900 which offered the first mathematical formulation of a random walk – Bachelier was a mathematician not a physicist, but his adviser was Henri Poincaré, the great mathematician and theoretical physicist; (iii) the American economist Harold Hotelling (born in 1895) was educated as an astronomer; (iv) in 1942 the theoretical physicist E. Majorana wrote a paper on the essential analogy between statistical laws in physics and in the social sciences (cited in Mantegna and Stanley 2000); (v) the American theoretical physicist Elliot Montroll (1916–1983) wrote a book (with Badger) about the analysis of speculative phenomena (1974); (vi) Benoît Mandelbrot (born in 1924), whose work on fractal geometry and chaos had a tremendous impact in many fields, was (like Allais) educated at the Ecole Polytechnique – throughout his active life he made significant contributions both in economics and in physics; (vii) the German theoretical physicist Wolfgang Weidlich (born in 1931) wrote a book on sociodynamics and published a paper in the renowned economic journal *Econometrica* (1983).

1.2 Assessment of pre-econophysics

A natural question is whether pre-econophysics was indeed econophysics in the same sense as it is understood in this book. The answer is a mixed one. Thus several pre-econophysicists (e.g. Allais, Walras, and Weidlich) disregarded the main prescription of physical research, which is the necessity of comparing theory and observation. This cannot be really surprising for someone like L. Walras whose contacts with physics remained confined to his education as an engineer, but it is more surprising for someone like W. Weidlich who was a renowned professional physicist. The case of M. Allais is even more puzzling for (as we already mentioned) he devoted several years to research in experimental physics; he also held firm methodological views about the necessity of confronting theory and observation as illustrated by the following excerpt (1968, my translation): "No matter how elegant a theory, if it cannot be confronted with empirical evidence it has no scientific value whatsoever." Any physicist would certainly agree with such a statement. Unfortunately M. Allais hardly followed that precept for his own economic research; in his main works one can hardly find any statistical comparison between theory and observation. For instance his theory of risk (1955, 1979) remains a completely theoretical construction, without any connection with empirical evidence. A clue to the solution of this puzzle should probably be attributed to the "Zeitgeist" of that time (see chapter 3): indeed the 1950s and 1960s were a time of rapid development of mathematical economics.

However, the contributions of researchers such as V. Pareto or B. Mandelbrot, were certainly masterly prefigurations of the kind of research econophysics tries to develop.

2 Institutional econophysics

The beginnings of institutional econophysics can be traced back to the moment when it became possible to publish economic papers in physical journals. One of the first economic papers to appear in a physical journal was probably "Levy walks and enhanced diffusion in Milan Stock-Exchange" by R. Mantegna which appeared in *Physica A* (1991). However, as shown in fig. 2.2(a) the real emergence of econophysics took place in the mid 1990s. A number of renowned physicists (fig. 2.2(b)) had an instrumental role in getting this policy approved by the editorial boards of journals, such as *Physica, The European Physical Journal*, or the *International Journal of Modern Physics*. In order to explain why this was a major landmark one must recall some characteristics of economic journals.

2.1 Idiosyncrasies of economic journals

Physical journals welcome all kinds of experimental observations, especially if they are puzzling and cannot be explained by existing theories. Recent illustrations are the numerous papers about avalanches in sandpiles. Economic journals, on the contrary, are very reluctant to publish empirical observations, especially when they have no clear interpretation within the existing theoretical framework.

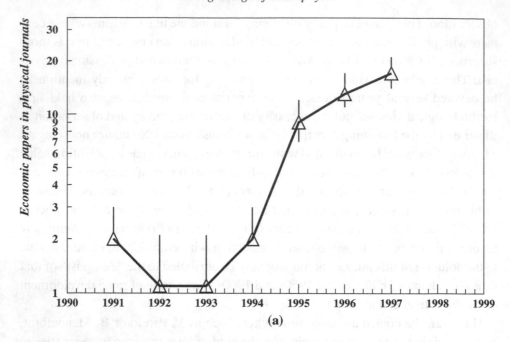

Fig. 2.2a. Number of economic papers published in physical journals in the first years of econophysics
Notes: Most of the papers were published in *Physica A*, *The European Physical Journal B*, and the *International Journal of Modern Physics*. For theoretical papers the distinction between economic and physical (or mathematical) papers is not always easy; the error bars reflect that uncertainty.

Another feature of great importance is the small volume of economic journals: the total number of pages published annually in physical journals is probably of the order of 10–20 times larger than for economic journals. As a result the acceptance ratios for papers in economic journals are between 5 percent and 25 percent, which in turn induces authors to write on the most fashionable topics in the hope of raising the interest of referees, rather than to pursue an original and consistent research project.

For all these reasons the fact that, from 1995 onward, physical journals welcomed economics studies was a watershed in the development of econophysics. It boosted empirical investigations and opened up new promising avenues of research. Unfortunately, for academic reasons few economists read physical journals or publish in them.

2.2 The beginnings of econophysics

Broadly speaking the four groups mentioned in fig. 2.2(b) were all working in statistical physics and were more or less closely in contact one with another. Needless

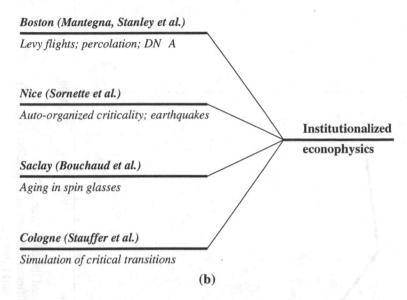

(b)

Fig. 2.2b. Emergence of institutionalized econophysics
Notes: The beginnings of institutionalized econophysics can be traced back to the moment when it became possible to publish economic papers in physical journals. The four groups mentioned in this figure were led by renowned physicists and their role was instrumental in getting that shift approved by the editorial boards of the journals. This list is certainly not exhaustive in the sense that several other groups contributed to this movement, such as the groups around D. Farmer (Santa Fe Institute), J. Feigenbaum and P. Freund (University of Chicago, Fermilab), S. Solomon (University of Jerusalem), Y.-C. Zhang (University of Fribourg, Switzerland), or the Olson group (Zurich).

to say the enumeration in fig. 2.2(b) is by no means exhaustive. Several other groups of statistical physicists took part in the movement, such as the teams around D. Farmer (Santa Fe Institute), M. Marsili and Y.-C. Zhang (University of Fribourg, Switzerland), P. Freund (University of Chicago), or S. Solomon (University of Jerusalem). One should also mention the Olson group (Zurich), a private financial institution which comprised several former theoretical physicists and was instrumental in making available to the community of econophysicists several huge data bases.

In retrospect one may think that the revolution of 1995 was not unexpected; it was so to say in the air. It is a common saying that revolutions are preceded by forerunners; that was also the case here as illustrated by the following episodes.

2.3 Neurophysics

In the late 1970s the study of networks of neurons became an important field of research in statistical physics. The objective was to propose theoretical models which could simulate various functions performed by the brain, such as

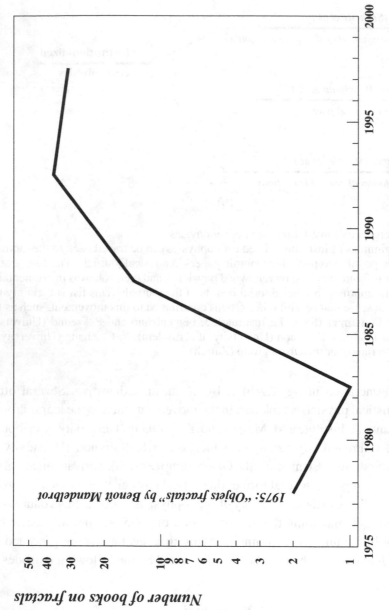

Fig. 2.3. Number of books published per half-decade whose title contained the term "fractals"
Notes: Between 1975 and 1985 the three books published were three different editions of Mandelbrot's (1975) book *Fractals*. These data are based on the electronic catalog of the Harvard Library.

long- or short-term memory. Unfortunately this research remained almost exclusively theoretical. Very few papers containing real data about biological neural networks were ever published in physical journals. The term "neurophysics" was apparently introduced in 1977 in a book published by A. Scott, but it has been used only rarely subsequently; there are only two other books in the Harvard Library catalog which contain that term in their titles – one published in 1981 and the other in 2000. Thus, the study of networks of neurons was rather a short-lived research bubble. Econophysics might meet the same fate if it restricts itself to building purely theoretical models. Fortunately it is easier to find statistical data for economic systems than for neural networks, and that will hopefully make the difference.

2.4 The fractal revolution

Benoît Mandelbrot's book entitled *Les objets fractals* appeared in 1975. It was translated into English two years later, but, as shown in fig. 2.3, no other book with a title referring to fractals was published in the half decade 1975–1979. In the half-decade 1980–1985 the only book listed in the catalog of Harvard Library was a new edition of Mandelbrot's book. In other words it took ten years for the concept of fractality to gain acceptance. In the late 1990s the notion of a fractal dimension became a tool commonly used in physics as well as in various other fields ranging from geology to biophysics and finance. In physics the concept fostered numerous experiments and simulations, but it did not bring about the formation of a new branch of physics; in that sense it is not comparable to econophysics.

2.5 Formation of an econophysical community

In the late 1990s several conferences and workshops took place which marked the beginnings of econophysics as a collective movement among the physical community. The main conferences are summarized in table 2.1. It can be noted that the Prague Workshop in February 2001 marked a turning point, for, in contrast to former conferences which were restricted to financial analysis, its title makes explicit reference to the broader field of economics; the proceedings can be found in Bouchaud *et al.* (2001).

For econophysics to survive the generation of the "founding fathers," and really become a subfield of physics, an essential step is the organization of courses in econophysics. At the time of writing several physics department in various countries offer courses in econophysics: examples are Duke University (Sergio Picozzi), Trinity College at Dublin (Peter Richmond), University of Munster (Jörg Lemm), University of Southern Denmark, and University of Ulm.

Table 2.1. *The beginnings of econophysics: conferences and workshops*

Date	Location	Title	Organized/sponsored by
July 1997	Budapest (Hungary)	Workshop on econophysics	
July 1999	Dublin (Ireland)	Application of physics in financial analysis	European Physical Society
July 2000	Liège (Belgium)	Application of physics in financial analysis	European Physical Society
Sept. 2000	Hefei (China)	Econophysics and financial complexity	Anhui Association of Science and Technology
Nov. 2000	Tokyo (Japan)	Empirical science of financial fluctuations	Physical Society of Japan
Feb. 2001	Prague (Czech Republic)	Application of physics in economic modelling	NATO

Notes: This list is not exhaustive in the sense that it does not include a number of smaller meetings which took place notably in Denmark and Germany. More details can be found on the web site: http://www.unifr.ch/econophysics. The city of Hefei is located in the province of Anhui, about 400 km to the west of Shanghai. Note that econophysics made a modest appearance (three papers presented under that heading) at the January 2001 conference of the American Economic Association (New Orleans).

2.6 A personal note

Before closing this part one should perhaps give a brief account of how the present writer was involved in the emergence of econophysics. After receiving a Ph.D in particle physics I continued to work in this field for some years, but in parallel I developed an ever-increasing interest in economics and history. In the early 1980s I realized that I could build a kind of "social physics" (to use the term introduced by Quételet in 1835), which would satisfy my interest for the social sciences as well as my dedication to physics.

From 1982 onward this endeavor materialized into a more or less continuous flow of papers; in order to suggest how eclectic my interests were at this time it may be worthwhile to cite the earlier ones. The first paper (1982a) was written at the Illinois Institute of Technology (Chicago), where I visited Professor Semyon Meerkov (a member of the brilliant Russian probability school who had just emigrated to the United States), and was devoted to the the sociology of hierarchical organizations; unfortunately it was a purely theoretical paper for I was unable to find empirical data which could be compared to the predictions of the model. The second paper

(Roehner and Wiese 1982) was about the Pareto law in urban dynamics; the third (1984) was a paper on macroeconomic growth. During the period 1982–1997 most of my papers were published in social science journals but, as already explained, this implied several limitations. In particular it was almost impossible to publish empirical studies, and in a methodological paper published in 1997 I advocated the creation of a new journal devoted to empirical and observational research. By a curious coincidence, at the very moment when that suggestion was made, it was both fulfilled and rendered needless; indeed, thanks to the emergence of econophysics, several physical journals began to welcome empirical economic papers.

My first contact with what was to become the econophysics community was through an email from Didier Sornette on 2 October 1995, and since that time we have had a permanent and stimulating collaboration. Other milestones were my encounters with Eugene Stanley in February 1998 and with Yi-Cheng Zhang in December 1998.

2.7 The future of econophysics

It is almost as difficult to predict the future of econophysics as it is to forecast the evolution of stock markets. As a matter of fact the two questions may well be related in some way. Institutionalized econophysics arose in the strong bull market of the mid 1990s and, not surprisingly at such a moment, its first topics of interest were in relation to financial markets. Fortunately, because there was no rigid compartmentalization in this new field, several econophysicists did not hesitate to shift from finance to economics and vice versa; see in this respect the papers by Plerou *et al.* (1999), Bouchaud and Mézard (2000b) or Solomon and Richmond (2001).

One may wonder what would be the consequences for econophysics of a bearish (or less bullish) stock market. In that respect one should bear in mind that financial markets are not the only places where collective speculative phenomena can be observed; there are even good reasons (see chapter 1) to think that the main features of speculative behavior may better be studied on simpler markets. In other words, if, as can be expected, the shift to a bear market weakens the interest in equity markets this should not altogether be a bad thing from a scientific perspective.

The incidence of equity markets on econophysics is an exogenous factor which is beyond the control of econophysicists. However, our previous discussion about biophysics and neurophysics highlighted the importance of a close connection between model building and observational research. The achievements of biophysics, geophysics, or astrophysics certainly owe a lot to the interaction these disciplines were able to establish between theory and observation. Economics is beset by an over-abundance of models and a lack of data with which they can be confronted. Therefore it would be a rather poor achievement for econophysics to restrict its contribution to a number of additional models. On the contrary, by developing a new approach to observation and empirical investigation, econophysics can really invigorate economic research.

Part II

How do markets work?

3

Social man versus homo economicus

During the twelfth and thirteenth centuries at least 50 cathedrals were built within the geographical limits of contemporary France. For the society of that time this represented an overpowering investment the magnitude of which was probably (in relative terms) comparable to the one required for the construction of the railroad network in the second half of the nineteenth century. Was it a rational investment? Obviously the answer very much depends on how economic rationality is defined. In the eyes of the people of the twelfth century it was undoubtedly the right decision, for no other activity was deemed more worthwhile than to celebrate the glory of God. For twentieth-century Americans or Europeans it is certainly difficult to imagine to what extent religion permeated the society of that time. Education, common law, intellectual life, government, every aspect of social life was deeply influenced by religion. For instance, about one half of all the books published in the seventeenth century were concerned with religious topics (Bendix 1978); although we do not have a similar statistic for the twelfth century it can be assumed that the proportion was even higher.

One might think that nowadays at last, the definition of "rational economic behavior" is fairly well established, but that is probably only because we lack the benefit of hindsight. It is no longer the celebration of the Glory of God which is our ultimate aspiration, but it is hardly more rational to consider that no expense should be spared for the preservation of health; after all, once we are dead, the people of the twelfth century would reason, do we not share the Glory of God?

By comparing two societies which are so different one from another our objective is to suggest that the way we think and reason depends upon the society in which we live. The rules of logic are the same of course, but the pre-conceptions, assumptions, and expectations which span and structure our reasonings are conditioned by our social environment. This, of course, is an idea which has become familiar to sociologists and historians, at least since the work of E. Durkheim, but for people who remain permanently immersed in their own society this can be easily forgotten.

As far as economic rationality is concerned there is no need to go back as far as the twelfth century in order to observe changes in pre-conceptions. In the period 1950–1970 people reasoned in the framework of the Keynesian (i.e. government-oriented) economy, which came into being in the wake of the Great Depression. Then, in the 1980s, as the generation who had experienced the Great Depression were replaced by people for whom it was just history, neoliberal ideas came back into favor, and "privatization" became the battle cry of the day.

Such shifts in conceptions are not confined to macroeconomic questions, they also affect microeconomic issues and in particular the attitude of investors and speculators. In short, we argue in this chapter, the concept of the homo-economicus should be replaced by a homo-sociologicus, that is to say a social man in close interdependence with the society in which he lives, works, and speculates. The decisions taken by the social man will be no less rational than those taken by the homo-economicus, but his rationality framework will be society dependent. In subsequent chapters it will be seen that such a change in perspective has far-reaching consequences for the understanding of speculative bubbles. For instance the fact that between 1998 and 2001 the development of e-commerce was far less rapid than expected by Internet retailers is certainly to be attributed as much to social factors as to economic reasons. In the course of time the weekly trip to the supermarket had become a familial and social tradition that people were not ready to drop overnight. Of course, if online retailers had been able to offer huge discounts the shift would have been faster; unfortunately the grocery business is characterized by high fixed costs and low profit margins, which precluded substantial productivity gains.

Awareness of the fact that economic rationality is society dependent could lead to the pessimistic conclusion that there can be no stable patterns. However, at the end of this chapter, we explain how invariant patterns can be identified in spite of a changing social environment. But first of all let us investigate in more detail how the social man is related to the "Zeitgeist."

1 The social man and the Zeitgeist

"Consciously or unconsciously [the American Negro] has been caught up by the Zeitgeist and with his black brothers in Africa, he is moving with a sense of great urgency toward the promised land of racial justice" (Carson 1999). This is how, in a letter written in Birmingham (Alabama) jail, Martin Luther King describes the struggle for civil rights; by Zeitgeist (a word of German origin which literally means the spirit of the time) King understood both the mood of the time and a kind of historical necessity. At another place in the same book he uses again the word to emphasize that a boycott would have taken place in Montgomery (Alabama) anyway, whether or not he had gone there. As a matter of fact it is certainly no coincidence that the climax of the struggle for civil rights in the United States occurred between 1957 and 1963, that is to say at the very moment when most African countries won their independence. It is certainly not obvious to imagine through which channels the events in Africa have affected the African–American community; that is precisely why the word "Zeitgeist" is useful, it summarizes a number of factors and mechanisms whose effects can be observed in an unambiguous way even though we do not understand them in detail.

The Zeitgeist plays a determinant role in shaping the way people understand economic rationality. This is illustrated by the following examples. Back in the eighteenth century the main objective of the great powers was to develop colonial monopolies and empires, and, not surprisingly, mercantilism was then seen as the only sound economic doctrine. An eighteenth-century mercantilist would be appalled and bewildered by a trade deficit amounting to 3.5 percent of the GDP, as was the case for the United States in 2000. In the wake of the crash of 1929, speculating on credit became viewed as highly hazardous and was therefore discouraged by all means. Nevertheless in the last decades of the twentieth century, boosting consumption and speculation by encouraging people to rely more and more heavily on credit was regarded as a perfectly acceptable policy.

In the following paragraph we analyze in more detail two microeconomic cases which will permit us to better understand how the Zeitgeist is influenced by the fast growth sectors of the economy. Typically the Zeitgeist ensures that no obstacles are left in the way of fast growth sectors, that may slow down their development.

1.1 Connection between fast growth sectors and Zeitgeist

1.1.1 Slave trade

In 1781, Luke Collingwood, captain of the slave ship Zong bound for Jamaica, ordered his crew to throw 130 sick African slaves overboard to economize on food and water and save the rest of his cargo. The following year the ship-owners in England claimed compensation (£30 per lost slave) from the insurer and took them to court when they refused to pay. The legal issue was not whether the captain had the right to drown the slaves – that right was not contested by the court – but only about the cargo and whether the captain had any reasonable alternative in that emergency situation (*Sunday Times*, July 5, 1992, Walvin 1992).

This example is interesting because it shows both permanence and changes in conceptions. Permanence, because ship-owners and insurers would argue in a very similar way today on a question about a (non-human) cargo. Changes, because nowadays it would be unthinkable for a court to apply such a reasoning to a cargo of human beings.

Let us now try to understand how the conceptions about the slave trade were influenced by the requirements of fast growth sectors. The height of the transatlantic slave trade was in the mid eighteenth century, and one should remember that this period was marked by fierce competition between the great powers for the conquest of colonies. The Seven Year War (1756–1763) between Britain (and her allies) and France (and her allies) was in fact a world war, and the stakes were the colonies of Canada, India, the Caribbean Islands, and Florida. The sugar plantations located in the Caribbean Islands were a rapidly developing sector which generated hand-some profits. Because of the climate the slaves in these Islands had only a short life expectation, which explains why they absorbed the largest part of the transatlantic slave trade. The momentum generated by this business was apparently sufficiently powerful to sweep away all other considerations. Whether Anglicans, Lutherans, Quakers, or Roman Catholics, all slave merchants were able to find in the Bible religious and moral justifications for their trade. Subsequently, in the nineteenth century, cane sugar was increasingly challenged by sugar beet, and the Caribbean plantations lost their importance. Not surprisingly, the Zeitgeist progressively be-came less favorable to the slave trade. In the late nineteenth century, putting an end to the residual slave trade even became a justification for colonial intervention in continental Africa.

Not surprisingly, the new wave of colonial expansion came up against a man-power problem for the development of plantations in the Fiji Islands, Sri Lanka, Kenya, Uganda, or Surinam, which had many similarities to the labor shortage encountered formerly in the Caribbeans. The solution which was adopted, namely indentured labor, consisted in transporting thousands of poor Indians to the plan-tations on long-term labor contracts. This system had many of the advantages of slave labor without its main (moral) drawbacks (for more details see Tinker 1974).

1.1.2 X-ray medical instruments

In the above example, thanks to the benefits of hindsight, we can make our point fairly easily. However, because it belongs to the distant past, the case is probably not completely convincing. We now consider a situation which belongs to the second half of the twentieth century. In this case the research was far more hazardous for we had to collect and compare scattered and sometimes contradictory data. The very fact that conclusive evidence was so difficult to obtain attests that the prevailing Zeitgeist did not favor a critical assessment. The present discussion will probably be just as controversial as a discussion of slave trade would have been in the 1770s.

More specifically our discussion concerns the policy of preventive X-ray exam-inations. The American Cancer Society (ACS) recommends a mammogram every two years, from ages 40 to 50, and annually after 50; this policy has become ac-cepted medical practice not only in the United States but in almost all Western countries. Thus, in 1994 in the United States, 61 percent of all women aged 40 and over have had a mammogram within the past two years (*Statistical Abstract of the United States*, 1999, 135). One might think that this policy corresponds to a carefully defined social optimum. Let us briefly examine the relevant data.

In the early 1990s a mammogram cost about 70 dollars. Furthermore it can be estimated that at the moment when a tumor becomes detectable it has been growing for at least eight to ten years on average. That estimate was obtained on the basis of a doubling time of cancer cells of the order of one hundred days. Within six years the tumor contains one million cells and is of the size of a pencil point; at that stage it cannot be detected by mammography. It is only after eight years when the tumor contains one billion cells that it becomes detectable; incidentally, it also becomes detectable by palpation about one year later. This explains why at the time when the tumor is detected cancer cells may already have lodged into other organs.

However, as one knows, X-rays are not completely harmless. Can one estimate the additional risk due to mammographic exams? Such an exam exposes a woman to a dose of radiation which is between ten and 20 times greater than the dose for a chest X-ray, that is to say at least 250 millirems. Note that back in the 1960s the radiation dose was much higher: a typical mammogram delivered as much as 10 to 20 rems (*Denver Post*, August 2, 1995). For the sake of comparison one can mention that the average radiation dose received from natural sources (the so-called natural background radiation which is due to cosmic rays and other natural sources) is in the United States of the order of 250 millirems a year or that a one-way flight from Boston to Los Angeles results in a radiation dose of about 10 millirems (table 3.1).

Table 3.1. *Order of magnitude of radiation doses*

Technique	Radiation dose [millirem]	Index (flight = 1)
X-ray absorptiometry	5	0.5
Flight from Boston to Los Angeles	10	1
Dental X-ray	15	1.5
Chest X-ray	25	2.5
Natural radiation background (US, yearly)	250	25
Mammogram	250 to 1,000	25 to 100
Full mouth dental X-ray	1,000	100
CT scan	200 to 2,000	20 to 200

Notes: X-ray absorptiometry is used to estimate bone density. CT scan means computed tomography scan. The figures given in the table give only orders of magnitude in the sense that the actual doses largely depend upon the device that is used. In the early 1960s a typical mammogram delivered a dose of 10,000–20,000 millirems. A rough estimate is that if all people in a population of one million are exposed to a dose of 1,000 millirems this will induce a fatal cancer in 1–10 people. CT scanners are being progressively replaced by positron emission tomography and magnetic resonance imaging equipment which deliver no X-rays; magnetic resonance instead uses strong magnetic fields of up to one tesla.
Sources: Los Angeles Times (December 20, 1989), *FDA Consumer* (October 1986), Information Access Company (September 1991).

At this point a brief word is required about units of measurement. Two units are commonly used: (i) the rad characterizes the amount of energy which is dissipated; 1 rad corresponds to a dissipated energy of 1 erg per gram that is to say 10^{-7} Joule per gram. (ii) The rem (which means roentgen equivalent man) characterizes the biological effects of a radiation; it is the product of the intensity of the radiation expressed in rad and the ionization efficiency of the radiation. For X-rays one can consider that the ionization efficiency is of the order of 1, whereas for neutrons, for instance, it would be larger than 1. In other words, for X-rays 1 rad is approximately equivalent to 1 rem (*Encyclopédie des Sciences et Techniques* 1973).

Now, according to the Nuclear Regulatory Commission, the number of fatal cancers from an exposure to 100 millirem per year for every individual in a population of one million is of the order of $10^6/290 = 3400$ (*Boston Globe*, December 14, 1990). Such a global result is fairly imprecise however, for it is well known that a given dose is more harmful when it is received by a young subject (for the obvious reason that the cancer has more time to develop). Moreover, it has been estimated (Gofman 1985) that for a woman who has followed ACS guidelines since the age of 40, the risk of developing a breast cancer in subsequent years is almost doubled as a result of the mammograph exams. As a matter of fact there was a worrying 180 percent increase in the incidence of breast cancer in the United States between 1969 and 1991, but nobody knows if the two facts are related. Naturally, once a cancer is detected through a mammogram this results in an improved survival chance, but no study seems to have been conducted which would permit to estimate the net benefit of the mammographs in terms of the five-year survival rate. In the 1980s and early 1990s the common opinion was that in order to assess the effects on women of medical imaging devices one would have to follow a population of 60 million women from age 35 until death, a study which would be far too expensive.

A similar problem exists with CT scanners, i.e. computed tomography scanners. In the early 1980s about 1.3 million scans per year were done in the United States (United Press International, November 7, 1981). The radiation dose is in the 300–500 millirems range or even higher (Townsend Letter for Doctors and Patients, November 1995) and according to a fairly conservative estimate this dose would result in about 800 additional cancer deaths (*The National Journal*, April 28, 1979). Are we really sure that the benefit compensates the risks resulting from exposure to such high doses of radiation? Although such a question would seem to be of cardinal importance, it has in fact received little attention.

Like sugar plantations in the seventeenth century, the medical instrument industry is a fast-growth sector, generating handsome profits. In the 1990s there were more than 20 companies that manufactured and sold mammographic X-ray machines and CT scanners, and some of them, such as General Electric, were very large companies. In the 1980s a CT scanner cost between one half and one million dollars. Needless to say, once such a machine has been bought, there is a strong incentive to use it extensively.

With respect to these matters one observes the same attitude as during a speculative bubble: the catchword is "full speed ahead" and nobody wants to question the soundness of the move. Even if the price–earnings ratio of a given market stands at 90, people will discover good reasons to make it climb even higher. The potency of the Zeitgeist can hardly be overestimated.

1.2 Quantitative measure of the role of the Zeitgeist

Changes of the Zeitgeist are particularly obvious when observed on a long-term scale. To many readers the example regarding slave trade probably appeared more illustrative than the one about X-rays for which we do not yet have the benefit of hindsight. However, if one can find a quantitative characterization of the Zeitgeist, its role can be seen more clearly, even for short-term effects. This is what we have tried to do in fig. 3.1 and table 3.2. The table and graph describe three variables: (i) the number of books published yearly where the title contains one or more of the three words – stocks, stock market, or speculation; (ii) the number of stock and bond salesmen in proportion to total population; (iii) the level of share prices.

Table 3.2. *Social manifestations of bull versus bear markets*

	1900	1910	1920	1930	1940	1950	1960	1970	1980	1990	1995
S. and P. 500 (deflated)	6.15	8.36	3.32	10.5	6.55	6.36	15.7	17.8	17.7	20.9	33.4
Books	2.6	9.0	10	19	10	7.8	9.0	17	17	18	40
Stock and bond salesmen (per 100,000 population)	4.76	5.88	9.32	15.8	12.9	7.14	15.8	47.8			

Notes: "S. and P. 500" refers to the Standard and Poor's index of 500 American common stocks. The books row refers to the number of books published annually for the five-year intervals centered on the given years, and whose titles contain one (or more) of the words "stocks," "stock market," "speculation"; these data are based on the Harvard Library online catalog. The correlation of the logarithms of book and stock price data is 0.38 (0.56 when one considers nominal stock prices); the correlation of the salesmen and stock price data is 0.68 (0.85 when one considers nominal stock prices). Between 1910 and 1940 the correlation between annual stock prices and number of books is 0.58.
Sources: Historical Statistics of the United States (1975), http://finance.yahoo.com

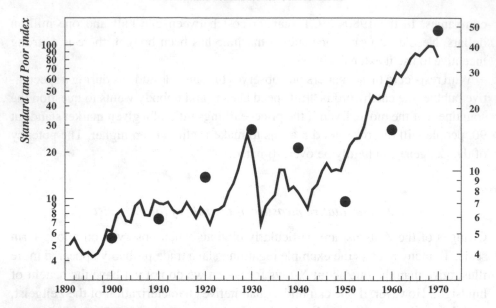

Fig. 3.1. Relationship between the level of stock prices and the number of stocks and bond salesmen
Notes: Solid line: Standard and Poor's index of common stocks (nominal prices). Dots: proportion of stock and bond salesmen in 100,000 population (right-hand scale). The correlation between number of salesmen and stock price levels at the same dates is 0.85. The number of salesmen is only one illustration among many others of how stock price fluctuations are interconnected with socio-economic swings; another example is the number of books about stocks published annually (see table 3.2).
Source: Historical Statistics of the United States (1975).

How should the fact that the variations are broadly parallel be interpreted? The relationship (correlation is 0.85) between the number of salesmen and the level of share prices could seem natural; however a little reflection shows that it is not quite so obvious. First of all it must be remembered that the salesmen sell bonds as well as stocks, and bonds are usually a good investment when the stock market is bearish. Moreover, it should in principle be possible to attract the interest of people in stocks even when prices are low; after all should one not, as the saying goes, buy at troughs and sell at peaks? As a matter of fact, the decrease in the number of salesmen during bear markets seems rather to show a marked disinterest by the general public for everything which is related to stock markets. That interpretation is confirmed by the evolution in the number of newly published books. In this case there is clearly no rationale (whether financial or intellectual) for a slowdown during bear markets. On the contrary, one would expect that economists would try to understand why the speculative peak came to an end. The simplest interpretation of the evidence in fig. 3.1 is that during bear markets the public (in which one should also include the economists) simply turns its back on stock markets. Similar tests based on newspaper articles point in the same direction.

1.3 Ways and means of the Zeitgeist

So far we have refrained from indicating by which specific mechanisms the Zeitgeist affects the behavior of economic agents. As a matter of illustration we consider here two possible mechanisms.

- Consider a trader in the monetary exchange market; let us assume that there is a strong trend in favor of an appreciation of the yen. Assume further that for some reason our trader has become convinced that the yen is now overvalued; is he (she) *ipso facto* going to sell yens and to buy dollars or euros? If he does and the market does not follow he will suffer great losses and face the likelihood of losing his job. No doubt he will include this consideration in his model, whether explicitly or implicitly. As a result the trader will tend to act in conformity with the prevailing trend, i.e. his behavior will be rational in the sense that it will take into account the behavior of other traders; this kind of rationality will subsequently be referred to as extended or collective rationality.
- Our second example concerns the expansion of European colonialism in Africa and Asia at the end of the nineteenth century. First it must be emphasized that these were areas of rapid development. For instance the profit rate of French colonial companies ranged in 1913 from 50 percent to 100 percent (or even more for some companies such as the Compagnie des Mines d'Ouasta for which it reached 125 percent). Between 1913 and 1929 the index of French stocks rose from 100 to 544 but for colonial companies the index climbed to 785 (*Les Collections de l'Histoire* 11, 67). In the early twentieth century colonial expansion was promoted by a number of societies among which the Union coloniale was the most influential; between 1894 and 1903 it spent one million francs for promotion and public relations purposes. After 1919 it distributed free papers twice a week to about 90 newspapers (*Les Collections de l'Histoire* 11, 31). One of the climaxes of the colonial Zeitgeist was the International Fair for Colonies and Overseas Dominions held in Paris in 1931, in which even the United States participated.
- Our third example is an another illustration of the same kind of mechanism. In the late 1990s Internet companies were undoubtedly fast-growth companies. As a result all obstacles that might hinder their development had to be set aside, the privacy requirement was certainly a crucial issue in this respect. In early 2001 the Online Privacy Alliance, which comprised major companies such as Microsoft, America Online-Time Warner, IBM, Sun Microsystems, organized a campaign in Washington with the purpose of stifling any Internet privacy legislation. It published the results of four industry-funded studies, asserting that privacy legislation would cost consumers billions of dollars annually (Reuters, March 13, 2001 and an online edition of the *Wall Street Journal* of the same day). In 1999 the United States Congress ruled that banks or credit card companies could share and sell information about consumers' spending habits, unless the latter expressed their opposition in writing. So intricate was that procedure, however, that by June 2001 fewer than 1 percent of consumers had exercised their right to protect their privacy. An article published in the newspaper *USA Today* (July 10, 2001) points out that the financial services industry had won clout in Congress during the 1990s with campaign contributions amounting to about 200 million dollars.

Dropping the old, narrow, and individualistic concept of rationality in favor of an extended and collective rationality could seem a daunting revision, but as we will see in the next section it is in line with the Paretian approach.[1]

2 Regularities

2.1 The search for uniformities and regularities

The central theme in the work of Vilfredo Pareto is the search for what he calls uniformities. For instance, by comparing income and wealth data in different countries he found that (at least in the upper range) their distribution $f(x)$ can be described by a power law: $f(x) = A/x^\alpha$. He used a similar approach in his sociological research. His major work is a 800 page catalog of uniformities, that is of recurrent attitudes in various societies. For example, the fear of having crops destroyed by hail or thunderstorms leads to magical or religious practices destined to keep away such calamities, which took similar forms in different societies. The search for uniformities and regularities (the latter word will be used preferentially when the uniformities can be stated in quantitative form) was in fact a major objective in sociology. Besides Pareto one should also mention the names of E. Durkheim (1858–1917) or Karl Deutsch (1912–1992), who studied social phenomena for their own sake without trying to derive them from individual motivation or behavior. Such an approach is particularly appealing to the physicist, who is very much aware of the fact that the relationship between individual and collective behavior can be complex and tricky to explain. One only needs to recall the non-trivial relationship between individual spins and the phenomenon of ferro-magnetism.

In contrast the cornerstone of most economic theories is the standard model of a rational (in a narrow sense) individual. This assumption has two far-reaching consequences: (i) a disinterest in collective phenomena that cannot be simply derived from the behavior of individual agents; (ii) a tendency to overlook the interactions between economic agents and the society in which they live. As a result an interdependence, such as the one suggested by fig. 3.1, would probably be disregarded as being of little interest. Several distinguished economists, such as H. Simon (1916–) or G. Snooks (1993), recognized this difficulty and proposed remedies. In the next paragraph the search for regularities is illustrated by various instances of speculative peaks. The figures have two main purposes: (i) they provide a first contact with the speculative phenomena which will be considered in more detail in subsequent chapters; (ii) they emphasize that the study of speculative phenomena should not be restricted to financial markets. As a matter of

---

[1] This idea has been rediscovered in the early 2000s by the behavioral economists; in the introduction to a collection of essays on behavioral economics, G. Lowenstein, C. Camerer, and M. Rabin wrote: "the strict rationality assumptions that many economists still embrace may someday be seen as a quaint, unrealistic, special case."

fact, commodities, land, real estate, postage stamps, and antiquarian books provide instances of speculative bubbles, which in many respects are simpler to study than those occurring in financial markets.

2.2 Examples of speculative peaks

2.2.1 Commodity markets

From a methodological point of view commodities have two great advantages: (i) The fact that there are many sorts of commodities (wheat, bananas, sugar, silver, copper, and so on), which differ in production, transportation, storage, or consumption characteristics, gives opportunity to study the role of these different parameters. For instance, if one wants to test a model of speculation, based on storage costs (e.g., Deaton and Laroque 1992), one can consider a sample of products characterized by a broad range of storage costs. (ii) Price statistics are easily available, for instance from Commodity Exchanges or from UNCTAD (United Nations Conference Board on Trade and Development) or IMF (International Monetary Fund) publications.

Fig. 3.2 shows two peaks for wheat prices in France during two six-year intervals, which are about one century apart. The amplitude of the peak (i.e. the ratio of peak price to initial price) or the duration of the up- and down-going phase are fairly

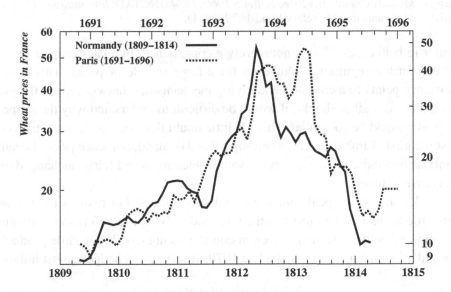

Fig. 3.2. Price peaks for wheat
Notes: Solid line: monthly prices in the departement of Seine-Inférieure (subsequently renamed Seine-Maritime, main city is Rouen) in Normandy; the prices are expressed in francs (left-hand and bottom scales). Dashed line: monthly prices in Paris expressed in livres (right-hand and top scales).
Sources: Labrousse, Romano, and Dreyfus (1970), Baulant and Meuvret (1960).

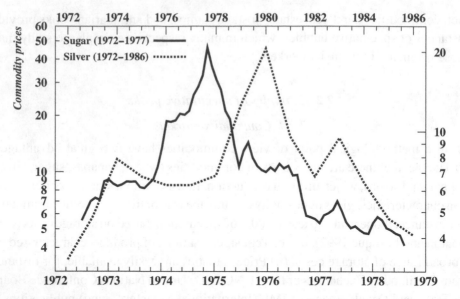

Fig. 3.3. Price peaks for sugar and silver
Notes: Solid line: monthly price of sugar in Caribbean ports expressed in US cents per pound deflated by the US consumer price index (left-hand and bottom scales). Dashed line: annual price of silver in New York expressed in US dollars per ounce troy (right-hand and top scales).
Sources: Monthly Commodity Price Bulletin 1960–1984 UNCTAD; *International Financial Statistics* (International Monetary Fund 1979, 1991).

similar in both cases. This is not merely a coincidence. In a later chapter we will see that such a regularity holds even for a large sample of peaks. This kind of invariance points to a common underlying mechanism; otherwise, given the great variability in weather shocks, it would be difficult to understand why the shape of the peaks would be so similar. There is little doubt that such peaks were triggered by some kind of imbalance between supply and demand, but, once prices began to climb, a speculative dynamic was activated which remained fairly unchanged over the course of time.

Fig. 3.3 shows two peaks for sugar and silver prices. For reasons not yet well understood, sugar is the most volatile commodity (volatility refers to the standard deviation of the price series). Silver, in contrast, is not a highly volatile product in 'normal' circumstances, but in the late 1970s it served as a hedge against inflation.

2.2.2 Land and real estate

The land and real estate markets are interesting for at least two reasons: (i) In modern economies these markets are among the largest in terms of capitalization. In most countries the capitalization of land and real estate markets are about two

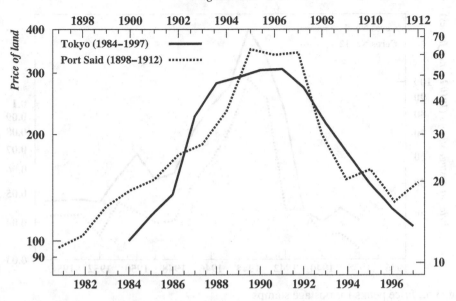

Fig. 3.4. Price peaks for land
Notes: Solid line: annual (nominal) prices of commercial land in the Tokyo area (left-hand and bottom scales). Dashed line: annual price of land in Port Said at the entrance of the Suez Canal expressed in French francs per square meter (right-hand and top scales). Although these episodes were almost one century apart, they were rather similar in the sense that in both cases there was a simultaneous stock market crash; between April 1907 and February 1913 stock prices dropped by about 50 percent.
Sources: Financial Times (October 17, 1997), Bourgeois (1913).

or three times larger than the capitalization of the equity and bond markets. As a result fluctuations in property prices can have stunning macroeconomic effects, as was demonstrated by the economic recession which resulted from the downturn in the Japanese property market in the early 1990s. (ii) Because they lack the sophisticated hedging tools (futures, options, and so on) that characterize equity and bond markets, property markets are much simpler.

Fig. 3.4 shows two peaks for land prices in time intervals that are almost one century apart. There is a great similarity between both cases, in the sense that they were accompanied by an abrupt decline in stock prices. Naturally the capitalization of stocks in 1900 Egypt was much smaller than in 1990 Japan, and had therefore only minor macroeconomic effects.

2.2.3 Postage stamps and antiquarian books

In spite of their negligible importance in terms of capitalization, the postage stamp and book markets in many respects provide an ideal laboratory for the study of speculative peaks. Let us see why. One of the main difficulties when studying

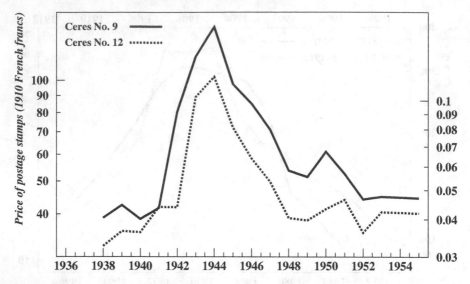

Fig. 3.5. Price peaks for postage stamps
Notes: Solid line: ten centimes French stamps issued in 1852–1853 (left-hand scale); the "Cérès number 9" indication refers to a commonly used stamp catalog; the prices are deflated and expressed in francs of 1910. Dashed line: five centime French stamp issued in 1853–1860 (right-hand scale). Whereas the number 9 is a rare and expensive stamp, the number 12 is common and inexpensive; nevertheless they experienced parallel price increases. Note however that the inexpensive stamp has a smaller peak amplitude than the costly one; this effect will be studied in more detail in a subsequent chapter.
Source: Massacrier (1978).

speculative peaks is to identify and control the exogenous shocks which trigger the bubbles. For the markets under consideration exogenous shocks are minimal. There are no weather shocks, no consumption shocks, and only a small interaction with other speculative markets. Moreover for stamps the "production" figures are well known; indeed for each newly issued stamp the total number of stamps offered to the public can be found in the stamp catalogs. The same remarks also apply to the market for antiquarian books, which is considered in more detail in a subsequent chapter.

Fig. 3.5 shows two peaks for stamp prices in France during World War II. As for the case of silver (fig. 3.3) stamps offered a hedge against inflation at a moment when, because of war restrictions, few other goods were available.

2.2.4 Share prices

In the early years of econophysics the study of stock markets monopolized the attention of many researchers. There was one obvious reason for this, namely the fact that huge data sets were easily available through the Internet. The question of stock price fluctuations is undoubtedly a fascinating problem but, for reasons

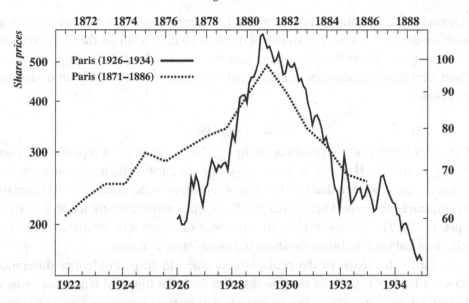

Fig. 3.6. Peaks for stock prices at the Paris stock exchange
Notes: Solid line: monthly share prices (left-hand and bottom scales). Dashed line: annual share prices (right-hand and top scales).
Sources: Hautcoeur (1997), White (1996).

we have already explained, it is also a very difficult problem. As far as the study of speculative peaks is concerned, one of the main obstacles is the fact that the number of major speculative episodes is rather limited. The number of cases for which detailed statistical data are available is of the order of 15, which makes the task of analyzing a phenomenon which depends upon a large number of factors with such a small sample an almost hopeless task. It amounts to understanding the mechanisms which govern the formation and movements of meteorological depressions from the observation of only 15 cases.

Fig. 3.6 shows two peaks for share prices in France. At first sight it could seem that these peaks are not very different from those in previous figures; however one should realize that, in contrast to previous cases, fig. 3.6 shows the evolution of an index, that is a weighted average of individual stock prices. As will be shown in subsequent chapters, depending on the sector to which they belong, individual stocks can show very different movements. In other words, in spite of the fascination they have held for researchers, it is not obvious that stock market downturns correspond to well-defined phenomena. In the nineteenth century, when railroad stocks represented over 50 percent of the capitalization of the NYSE, a market index had a well-defined meaning. Nowadays the situation is more complex, as different indexes can exhibit quite contrasting tendencies. Once again a meteorological parallel can help to illustrate that point. The level of the river Rhine depends on a great number of factors: if in spring the temperature jumps from 0 to 15 degrees (Celsius) within a few days, in addition the winter was very snowy in the Alps, and in addition the compensating reservoirs are almost full, then there can be a major flood. But trying to understand why these circumstances may occur simultaneously is largely an ill-defined question.

2.2.5 Conclusion

The speculative peaks represented in figs. 3.2–3.6 could seem repetitiously (and boringly) similar. However it is precisely this similarity which is of interest. For instance, it can be seen that the duration of a peak is of the order of 5–10 years for an amplitude comprised between 3 and 5 (except for sugar where the amplitude is equal to 10). This means that the price increase rate cannot be arbitrarily large and this gives valuable information about the speculative dynamic.

As far as the shape of the peaks is concerned there is an obvious difference between land or real estate price peaks with their flat tops and other cases which display rather sharp peaks. This difference will be stated quantitatively in a subsequent chapter.

4
Organization of speculative markets

François Castanier, a French ship-owner from the southern city of Montpellier was one of the six directors of the "Compagnie d'Occident," the company initiated by John Law. At the beginning of 1719 he wrote numerous letters to several middlemen in order to convert his shares in the company into land and estates. On September 18, 1719 he noted in his diary that, due to the speculation set into motion by Law's companies, he had few competitors for his purchases of land. F. Castanier thus provides an early example of those cautious investors who are able to withdraw from speculative operations in good conditions. Castanier's biography (Chaussinand-Nogaret 1970) contains several other elements which are of interest for the reader of the twenty-first century. At one point we are told that the nephew of one of his colleagues, Marquet de Bourgade, had spent several months in Amsterdam, where he familiarized himself with the so-called privilege contracts. Through such a contract one party acquired the right, but was not thereby obligated, to buy from (or sell to) the other party a given amount of a commodity at a certain price. In other words privilege contracts were what we today refer to as call (or put) options.

From these observations it can be seen that the behavior of investors as well as their financial techniques were not altogether different from those in use three centuries later. This will be one of the main themes in this and the next chapter. In pursuing this historical review we will try to discover some basic principles which guide the development of speculative markets. We will see that finance has been subject to the same trends as other industries, in the sense that product standardization, increased concentration, which allows higher trading volume, and avoidance of ruthless price wars have played a crucial role.

We describe and compare the organization of stock exchanges for different countries and time intervals. It will be seen that market rules often have had a dramatic impact on the occurrence and unfolding of speculation and panics. We repeatedly stress the importance of such collective constraints; once they are taken into account it often turns out that there remains little room for individual optimization. Next we briefly discuss the main trading techniques, e.g. short sales, buying on margin, and options, and we show how they are related to one another. Finally the fourth section provides long-term time series for such important variables as share indexes and bankruptcy rates.

1 Trends

Equity and to a lesser degree commodity markets stand out among speculative markets by the sophistication of their organization; it is for that reason that the present chapter is mainly devoted to the description of stock markets.

1.1 Concentration

Currently in the United States stock markets are concentrated in New York, while the options and futures markets are largely concentrated in Chicago. One is so used to this situation that one tends to forget that as recently as 1900 more than one hundred exchanges were functioning across the United States (Blume, Siegel, and Rottenberg 1993).

1.1.1 Role of declining communication costs

In 1932 a five minute phone call from New York to Chicago cost 5 dollars and 15 dollars to San Francisco. When estimated in terms of number of words transmitted the cost of a teletype transmission was almost the same as that of a phone call. The corresponding figures are summarized in table 4.1. How do such expenses compare with other transaction costs? In the 1930s the average number of shares per trade was much smaller than by the end of the century; a trade of 100 shares

Table 4.1. *Telephone rates*

	Telephone 2 minutes: 200 words	Teletype 5 minutes: 200 words
1932　(United States)		
New York to Chicago	$ 2	$ 2.2
New York to New Orleans	$ 3	$ 3.0
New York to San Francisco	$ 6	$ 4.8
2000　(France)		
Brest to Strasbourg	$ 0.26	
Paris to New York	$ 0.44	

Notes: The figures given for 2000 are averages over various rates.
Sources: *Fortune* (March 1932, 108), France Telecom.

could be considered as an average trade. If one takes 10 dollars as the average price of a share and 0.5 percent as an estimate for the commission received by the broker, the commission for a 100 share trade was 5 dollars. Thus, we see that the cost of the phone call was of the same order of magnitude as the commission. Such a cost was even more prohibitive for the short-term trades carried out by brokers, for in this case the margin was often very small and in order for the transaction to be profitable transactions costs had to be as low as possible. In the light of this reasoning, it becomes obvious that regional stock exchanges fulfilled a useful economic role. One might think that in the early twenty-first century communication costs have become negligible; but this is not completely obvious, for what really matters is the ratio between the commission (fig. 4.1(a)) and the communication cost (table 4.1).

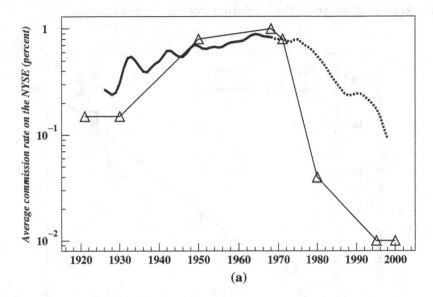

(a)

Fig. 4.1a. Average commission rate on the NYSE
Notes: Prior to 1968, both curves are based on minimum commission rates fixed by exchange rules; the thick solid line is based on annual data while the thin solid line is based on data for a number of sample years. After 1968 the thin solid line is based on commission rates announced by various broker firms; the thick dotted line is based on annual commission income reports by NYSE members. However these figures refer to bonds as well as shares and had therefore to be calibrated. Even if one takes into account a reasonable error margin in the calibration it seems obvious that there is a huge discrepancy between advertised rates and (actual) rates estimated from commission incomes. In a subsequent chapter we will see that this discrepancy can probably be explained by the difference between ask and bid prices (the so-called spread) which goes to the specialist. Note also that the transaction costs for mutual equity fund shareholders are markedly higher; in the mid 1990s they were of the order of 1.5 percent.
Sources: Thick line: adapted from Jones (2000); thin line: adapted from Roehner (2001a); equity funds: Reid (2000). I would like to express my gratitude to Richard Sylla (Stern School of Business) and Charles Jones (Columbia University) for their advice and assistance.

Recent developments have led to a situation where it is now just as convenient for an investor in Los Angeles to trade at the New York Stock Exchange as it is at the Pacific Stock Exchange located in San Francisco. Nevertheless, even after a century of concentration, there are still six regional stock exchanges, namely Boston, Cincinnati, Denver, Philadelphia, Salt Lake City, and San Francisco, but altogether they account for less than 5 percent of trading volume.

1.1.2 Productivity increase

As the concentration process unfolded, trading volume on the NYSE increased steadily (fig. 4.1(b),(c)) and the handling of such increasing volumes required huge productivity increases, which in turn implied drastic changes in the organization of the market. Let us illustrate this evolution through some examples.

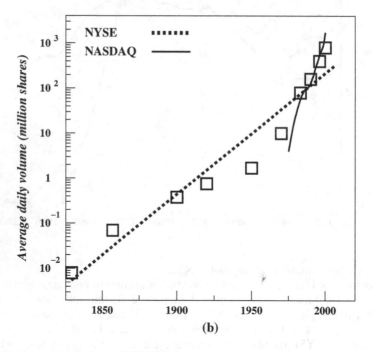

(b)

Fig. 4.1b. Daily trading volume 1829–2000
Notes: The squares correspond to data points for the New York Stock Exchange. The overall trend (dashed line) is an exponential increase with a doubling time of 11.4 years. However there have been long-term fluctuations around that trend; the trading volume was above its trend in the second half of the nineteenth century, lagged behind the trend in the first half of the twentieth century, and it largely outstripped the trend in the last decades of the twentieth century. The thin line corresponds to the NASDAQ market (created in February 1971); the annual growth rate is about three times as large as the long-term increase rate of the NYSE.
Sources: Blume, Siegel and Rottenberg (1993), *Statistical Abstract of the United States* (various years), http://www.marketdata.nasdaq.com/asp/masharevol.asp

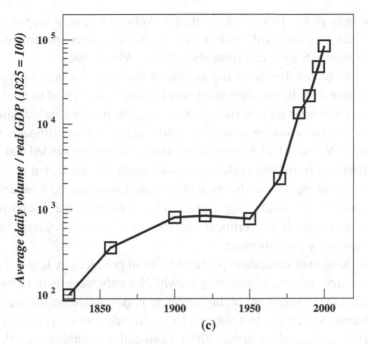

Fig. 4.1c. Increase in average daily trading volume on the NYSE with respect to GDP growth 1829–2000
Notes: In some time periods the increase in trading volume barely matched the growth of the economy, while in others it outpaced it markedly. Note that there is a long-term connection between stock valuation and trading volume. A high trading volume improves market liquidity and thus contributes to attract investors; in that sense it can be said that trading volume "oils the wheels of the market."
Sources: See previous figure.

First one should mention a major difference between US and European exchanges. On the former the positions of brokers were balanced at the end of each business day, whereas on most European markets this was done only every month or fortnight. Needless to say the first procedure was more demanding in terms of paper work.

As one knows at the NYSE every trade had to be registered by the "stock ticker," a machine generating a ribbon of paper on which all relevant information was printed. Such a device was first introduced in 1867. During the panic days of October 21–25, 1929, the daily volume rose to 12.8 million shares (October 24), which was about three times the average daily volume for September (4.2 million). Not surprisingly, the ticker began to run behind, a circumstance which prevented investors from getting timely information about the situation on the market and, therefore, contributed to increased panic. In order to avert such problems in the future a faster ticker (500 characters per minute) was installed in 1930. However, in the wake of the Depression, there was a marked decline in the trading volume,

and it was only in the 1960s that the trading volume began to surpass the record
levels of 1929. In 1964 a still faster machine (900 characters per minute) was put
in service (Blume, Siegel, and Rottenberg 1993, White 1990).

By the end of the 1960s, at the height of the post-war bull market, another
bottleneck appeared. In principle every stockholder was entitled to receive a stock
certificate. However as trading volume increased, the printing and sending of these
certificates became a burdensome task. Eliminating stock certificates was not a
simple matter. Not only had American stockholders to be persuaded that a monthly
holding statement from their broker was an adequate substitute, but it also brought
about a number of legal issues because many state laws required fiduciaries to hold
the certificates of the stocks they owned. The creation of the "Depository Trust
Company," which holds the certificates in the name of stock holders, was a major
step in productivity improvement.

The introduction of computers permitted a giant productivity leap but the transition to computers was made only very slowly. Not only were there many technical
problems, but there also arose highly sensitive issues regarding the role of market
makers. Blume, Siegel, and Rottenberg (1993) provide a vivid account of the issues
that faced the market makers in the 1970s (the so-called specialists) and the attempts
by the NYSE who tried to encourage the introduction of computers. The following
episode illustrates the climate which prevailed at that time. In 1971 one of the first
computer screens ever installed at the NYSE was put on a wooden platform two
feet above the specialists' post so that the prices on the display could be seen by
everybody. When Alan Loss, the president of the Exchange, came to his office on
one Monday morning he was surprised to see the computer lying on the floor in a
pile of sawdust; over the week end someone had used a saw to cut a semi-circle
through the wooden platform!

1.2 The thorny question of commission rates

"We will not buy or sell from this date for any person whatsoever any kind of public
stock at a less rate than $1/4$ percent commission on the specie value." This excerpt
from the Charter of May 17, 1792, which formulated some rules for the organization
of what was to become the NYSE, shows that the question of commission rates was a
central issue right from the beginning. A natural question is whether $1/4$ percent was
at that time considered as a high or a low rate. The book by Chaussinand-Nogaret,
mentioned at the beginning of the chapter, gives us an indication. In 1718 for a
purchase of six shares representing a total amount of 28,710 livres the commission
represented 12 livres, that is 0.04 percent or three times less than the 0.125 percent
mentioned in the Charter. Although one cannot draw definite conclusions from just
one comparison, it nevertheless suggests that a rate of 0.125 percent was probably
above average; had it been substantially below average it would have been pointless
to mention it explicitly in the Charter.

As a matter of fact it is not easy to form a clear picture of the commission actually charged by brokers on the NYSE. The Exchange's rules set minima, but there was usually a fairly wide margin between these minima and the actual rates. This point can be illustrated by the following example drawn from the very helpful reference book by J. Meeker (1922). In the early 1920s the commission for the purchase of 200 Southern Pacific (a railroad company) shares at 90 dollars represented 30 dollars, that is to say $30/(200 \times 90) = 0.17$ percent (the account sheet of that transaction is reproduced in the book). At that time the minimum set by the Exchange's rules was "2.5 cents per share for a stock selling between 10 and 125 dollars"; the application of this rule would result in a commission of $0.025 \times 200 = 5$ dollars, that is six times less than the actual rate.

The Exchange's rules also implied that no quantity discount was allowed, which means that the commission was a fixed percentage of the value of the shares whether the trade involved 100 or 10,000 shares. Since the work was almost the same in both cases this represented a substantial bonus for the brokers.The so-called "fixed commission era" lasted about 200 years and came to an end in May 1975 under the pressure of institutional investors (investment funds, pension funds, and so on). The role of these institutions steadily increased after 1960 as is apparent from the trend in the number of shares per trade (table 4.2): the proportion of small transactions has decreased steadily while at the same time there has been a substantial increase in the proportion of very large transactions. In the early 1970s, confrontation between the interests of the mutual funds and the rigid rules of the Exchange led to a fairly chaotic period during which the fixed commission policy, although still upheld in theory, was actually counterbalanced and almost nullified by the granting of various compensations to big clients.

It should be kept in mind that the commission is but one component of the total transaction cost: the cost of the communication (whether by phone or by computer) between the investor and the broker is an additional (albeit negligible now except for small orders) component. Another fairly small component is represented by

Table 4.2. *Small versus large transactions at the New York Stock Exchange*

	1920	1965	1970	1975	1980	1985	1990
Less than 100 shares	25%		5.9%	3.0%	1.8%	0.90%	0.71%
Over 10,000 shares		3.1%		16%	29%	51%	

Notes: The decrease in small transactions (the so-called odd lots) and the simultaneous increase in large transactions (the so-called block transactions) reflect the growing importance of institutional investors that is to say investment funds, pension funds, insurance companies.
Sources: Statistical Abstract of the United States (various years), Meeker (1922), Sobel (1987).

taxes. A more sizeable component is due to the bid–ask spread. Usually an order cannot be executed exactly at the price expected by the investor; it is precisely the role of specialists to keep the bid–ask range as small as possible, but in a highly volatile market there can be a substantial difference between the price expected by the investor and the actual price at which the transaction is carried out. Between 1930 and 1990 that component represented between 0.26 and 0.43 percent for a one-way transaction (i.e. a purchase or a sale); it dropped sharply in the decade 1990–1999 and was of the order of 0.1 percent by 1998 (Jones 2000).

Another practical problem encountered by small investors is mentioned in an article published in late 2000 by the *Wall Street Journal* (November 17, 2000): an investor may pay a broker a commission of one cent per share on an order to buy 1,000 shares (a small order) but end up paying 12.5 cents per share because the broker sent the order to a Wall Street intermediary that executed the order at a much higher price.

2 Trading techniques

The Employee Retirement Income Security Act (ERISA) was passed by the American Congress in 1974. It redefined the duties of corporate pension funds, stressing the performances of an entire portfolio as opposed to individual hold-ings. It had far-reaching consequences for equity markets for it permitted (and even invited) pension funds to invest some of their funds in shares. A similar move occurred at about the same time for insurance companies. In the following years, pension funds and insurance companies generated a considerable flow of funds toward the equity market.

One may wonder what made such a shift possible. Is it not the first duty of pension funds to hold safely the funds entrusted to them by depositors? At first sight it is not obvious that the best way to do so is to invest a sizeable proportion of these funds in stocks. Perhaps it is not entirely a coincidence that the Chicago Board Option Exchange opened in April 1973. As a matter of fact, resorting to such hedging instruments provided easier and more efficient means to secure portfolios against sudden falls in share prices. In short, better hedging instruments attract more investors, thus increasing trading volume, which in turn permits to lower commission rates. The interrelated role of these three factors is schematized in fig. 4.2.

In the following we first discuss hedging techniques available on modern markets with the purpose of emphasizing their common features and technical similarities, then we explain why in the course of time the new hedging instruments can also become a source of instability.

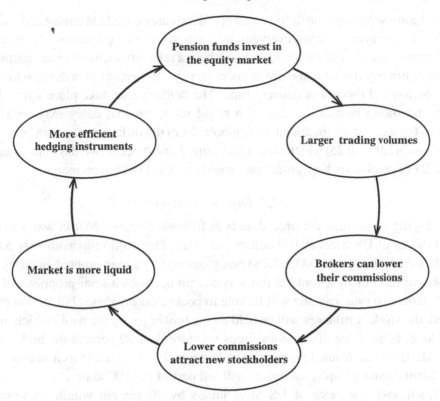

Fig. 4.2. Synergetic mechanisms driving the development of stock markets
Notes: The increase in trading volume is a key factor; it improves the market's liquidity which in turn tends to attract more investors. During the Great Depression the reverse mechanism took place: shrinking volume compelled brokers to raise their commissions which turned away investors.

2.1 Short selling, futures, options

2.1.1 A "not so simple" transaction

Suppose an investor (let us call him Martin) buys through the Internet 100 Cisco shares at 60 dollars per share, which he pays for at once by transferring 6,015 dollars (6,000 dollars plus a 15 dollars brokerage commission) from his account to the broker's account. This could seem to be a simple transaction, because it is similar to buying a dozen eggs in a grocery, but in fact it comprises several distinct operations which it may be helpful to list explicitly. In chronological order there are the following steps: (i) Martin sends the order to the broker; (ii) the broker tells Martin the price at which the transaction has been carried out – that price is of course close to, but usually slightly different from, the latest market price; (iii) Martin makes the payment; and (iv) the broker transfers the 100 Cisco shares to Martin's portfolio.

In the above transaction the different operations are carried out almost at the same moment; however, in many everyday transactions, these operations take place at different moments. For instance, if Martin buys a refrigerator, a car, or an apartment he does not pay the total amount at once; he pays 10 percent in order to settle the transaction and the rest at delivery time. The delivery may take place a few days after the transaction in the case of a refrigerator, but that delay may be a few weeks for a car or an apartment. In symbolic form the different steps can be noted: $1 - 2 - 3(2d) - 4(2d)$ which means that steps 3 and 4 occur two days after steps 1 and 2. For stocks, such a transaction, would be called buying on margin.

2.1.2 Buying on margin

For buying on margin, the procedure is as follows. Suppose Martin wants to buy 100 shares of US Steel at 100 dollars per share. He can do this with only 5,000 dollars cash, the rest (5,000 dollars) being covered by a loan granted either by the broker himself or by a bank. In this way Martin becomes a semi-proprietor of the 100 shares; on one hand, he will be able to pocket the dividends but, on the other hand, the stock certificates will be held by the broker (or by the bank) which made the loan. If the price of the stock drops by more than 50 percent the broker will ask Martin for additional cash, an operation which is referred to as a margin call; if Martin cannot comply, the broker will sell part of the 100 shares.

If, suppose, the price of US Steel jumps by 20 percent within six months, Martin will sell his 100 shares for a total amount of 12,000 dollars; he will then repay the 5,000 dollar loan, plus the interest, plus the commission, and keep the rest, that is almost 7,000 dollars. His profit rate on this operation will be $(7,000 - 5,000)/5,000 = 40$ percent in six months or 80 percent on an annual basis. If Martin had bought the stocks cash his profit rate would have been only $(12,000 - 10,000)/10,000 = 20$ percent. In other words, buying on margin allows higher profits.

In the 1920s the margin requirement was 25 percent on the NYSE, which means that an investor could buy stocks worth 10,000 dollars with an amount in cash of only 2,500 dollars. Since 1974 the margin requirement is 50 percent. On some markets, for instance for the purchase of commodity contracts on the Chicago Board of Trade, the margin requirement is only 10 percent.

The mechanism of options to which we turn now is merely an extension of the previous procedure.

2.1.3 Options

Suppose that the price of US Steel is 100 dollars per share and that for some reason Martin thinks this price will go up. He will buy the right to buy 100 shares at the price of 100 dollars per share at any time within the next six months. For this right he will pay 5 dollars per share. If the price of US Steel indeed goes up 20 percent in the next six months Martin will buy 100 shares for a price of 100 dollars per share at a moment when they actually cost 120 dollars. By selling them immediately after having bought them he will make a profit of $20 - 5 = 15$ dollars per share, which corresponds to a hefty profit rate of $15/5 = 300$ percent over a six-month period, that is 600 percent on an annual basis.

As one can see the mechanism is almost exactly the same as in the previous paragraph. However, options have two major advantages. First, the procedure is simpler: there is no need to secure a loan and, for the broker, there is no need to monitor the account of each client and to issue margin calls when required. In other words the options have a definite advantage in productivity terms. The second advantage, as shown by the above example, is the fact that the profit rate can be much larger, which is, of course, an attractive feature for investors.

2.1.4 How options can be used for hedging purposes

Suppose Martin has bought a large amount of US Steel shares at 100 dollars per share. In order to protect that investment against a brutal fall in the price of US Steel there are several possibilities:

- Martin can place what is called a "stop loss" order with his broker, which means that the broker should sell if the stock falls to the price of 90 dollars, for instance. This will not afford complete protection however: first, even if the "stop loss" order works, Martin will lose 10 percent; second, in some cases the price falls so quickly (examples are given in a subsequent chapter) that the order at 90 dollars cannot be executed at that price, in which case Martin can lose more than 10 percent.
- Another possibility is to purchase (at 5 dollars per share) the right to sell the stock at 100 dollars at any time within the next six months. If the price of the stock does not fall, the right to sell will not be exercised, and, in this case, Martin will lose 5 dollars per share. However, if the stock falls under 100 dollars, the right to sell will be exercised. Suppose the price of the stock has dropped overnight from 100 to 80 dollars, by exercising his right to sell Martin will benefit from much better protection than through a "stop loss" order.
- Another possibility is short selling. This means that Martin will borrow US Steel shares from a broker for a fee. He will then sell them with the intention of repurchasing them later and returning them to the lender. If the price falls he will be able to repurchase them at a lower price and thus make a profit which will balance his loss on his own US Steel shares. However this technique can become very perilous if the stock climbs markedly, for in this case Martin will have to repurchase the shares at a much higher price. He may suffer a great loss or may even be unable to repurchase the shares at that higher price, in which case he will face bankruptcy. In the nineteenth century there were many "corner-battles" which were based on that mechanism; in the twentieth century, because of the high risks involved, short-selling progressively dropped out of favor.

What should one retain from the above discussion? First, we have seen that if one leaves out technical terms (put-options, call-options, strike price, and so on) these trading techniques are fairly simple and bear much resemblance to each other. For the sake of brevity we did not discuss a number of other techniques, such as options on indexes or index futures, but they are based on similar principles. The second important idea is that, thanks to these hedging techniques, portfolio management can, at least in normal times, be optimized in a fairly effective way. Whether they will work as well in a prolonged bear market is another question.

The last decades of the twentieth century were marked by a bewildering multiplication of trading techniques and one may wonder why. This issue is considered in the next paragraph.

2.2 How to create a successful financial product?

The basic principles on which options, futures, or index futures are based have been known for a long time. As we mentioned at the beginning of this chapter, options were already used at the Amsterdam exchange in the eighteenth century. In the United States put- and call-options were used in the late nineteenth century at the Chicago Board of Trade (Emery 1896). They were also in use at about the same time at the Paris exchange under the name "marchés à primes" or at the Berlin exchange under the name "Prämiegeschäft." They were fairly popular during the bull market of the 1920s, but fell into oblivion during the subsequent bear market.

This leads us to the question raised in the title of this paragraph, namely: what makes the success of a financial product? There are at least five important factors: (i) high trading volume, (ii) standardization, (iii) high profit opportunity, (iv) backing of an important exchange, and (v) to be in line with trading habits.

In a general way high trading volume ensures the liquidity of the market, that is the ability of traders to find buyers and sellers with minimum delay.

Standardization has the same objective. Thus, in futures contracts for commodities, the quantity, quality, delivery date, and delivery point are all standardized, leaving only the price to be established by the contracting parties. In the same way, options are standardized: uniform expiry dates (at closing time on each third Friday), standardized strike prices (i.e. the price at which the shares can be bought or sold) at 5 dollar intervals for stocks under 100 dollars and at 10 dollar intervals for stocks over 100, uniformly increasing duration by a period of three months, and so on.

The possibility of high profit rates is of course a necessary condition in order to attract speculators and thus ensure high trading volume. As we will see in the next paragraph it is a common rule that a financial product that has been introduced for hedging purposes soon becomes itself a speculative target.

The third condition is closely related to the first in the sense that only major exchanges have large daily volumes and therefore good liquidity. This can be illustrated with the case of the options market. We have mentioned that in the wake of the Great Depression options fell into oblivion, but this is not completely true. In fact options were still traded under the patronage of the "Put and Call Brokers and Dealers Association," which regrouped a handful of dealers. As a result the options remained a narrow market with poor liquidity, a situation which even the bull market of the 1960s did not change. The real start of the options market came in 1973 when the Chicago Board Options Exchange (CBOE) opened. As we have seen, futures and options for commodities had been traded for almost a century at the Chicago Board of Trade and the standardized options traded at the CBOE were largely modeled on the contracts for commodities. In the 1980s options markets opened on other exchanges as well; thus an options market was launched at the Paris exchange in 1987. Worldwide there was a 12-fold increase in the trading volume of derivatives during the 1990s (*The Economist*, November 27, 1999, 103).

The last condition emphasizes the role of tradition. In order to explain what we mean let us consider the case of the rice market. A world production of about 400 million tons (in 1985) of rice makes it a commodity almost as important as wheat (world production was 520 million tons in 1985). Yet, in the mid 1980s there was no futures market for rice, whereas for wheat forward and futures markets had been in existence for more than a century (Mouton and Chalmin 1985): different continents, different traditions. This is in line with the so-called QWERTY effect conceptualized by B. Arthur: in short, present-day keyboards are as they are because once people got used to a given design it became difficult and costly to change it. Another example is provided by diamonds. Recurrent attempts in the 1970s and 1980s to create a futures market for diamonds were unsuccessful. One of the first attempts took place on the Pacific Stock Exchange (San Francisco) in the 1970s; in the late 1980s other attempts were made at the London Commodity Exchange and the New York Commodity Exchange (Boyajian 1988). Perhaps the existence of a futures market was simply incompatible with De Beer's quasi-monopoly.

There is another condition for the creation of a successful futures or options market that we did not yet mention, namely substantial price volatility. For a commodity (e.g. bananas) whose price remains almost unchanged over the course of time there is no need for hedging, and no profit opportunity for speculators, hence there is no justification for futures or options. In short, too much volatility is not desirable because it makes economic operations more difficult, but too little volatility would undermine the *raison d'etre* of financial markets. In the next paragraph we consider the question of the stabilization of financial markets in more detail.

2.3 Protection against market crashes

The main rationale for the introduction of the trading techniques discussed in the previous paragraphs is that they provide protection against severe market fluctuations. But these techniques are not the only means by which to ensure such protection. In the present section, we review more or less in chronological order some other stabilization means.

2.3.1 The specialist system

One of the main pillars of market stabilization is the specialist system. Specialists are members of the Exchange who provide an interface between the Exchange and the investors. Each specialist is in charge of one or two shares; they act as brokers but they can also buy and sell on their own account.

Specialists are also called market makers, an expression which can be understood in the following way. There are moments when everybody shares the same view about a given company, for instance after an announcement regarding higher (or lower) than expected earnings. In such moments it would be very difficult to find someone willing to sell (or to buy) and it is precisely the role of the specialist to maintain an orderly market and to prevent price fluctuations that are too sharp. For instance on the American Stock Exchange (AMEX), thanks to the participation of the specialists, the average difference between two successive sales of a given stock was 6.5 cents, which means that there were no abrupt variations (remember that the average price of a stock is around 50 dollars) as a result of a lack of buyers or sellers (Bruchey 1991). In compensation for their role, specialists are granted various prerogatives. For instance, they have sole access to order book information; in contrast, an investor cannot know the real number of shares available at the quoted price. They also play an important role at the beginning of the trading day by determining the price which will balance the buy and sell orders that accumulated during the night. On some occasions that equilibrium price can be very different from the closing price of the previous day.

During the bull market of the 1990s the intervention power of the market makers diminished: in 1987 they held 25 billion dollars in equity capital which represented 1 percent of the total value of US stocks; in 1999 their equity capital had tripled but, due to the rapid increase in stock capitalization, it represented only 0.4 percent of the value of US stocks (*New York Times*, April 6, 2000).

2.3.2 Investment funds

The main rationale for the introduction of investment funds was to provide greater security by dividing risks. This is expressed very clearly in a declaration made by Irwing Fisher, the renowned American economist, in September 1929: "A few years ago people were as much afraid of common stocks as they were of a red-hot poker. Why? Mainly because the average investor could afford to invest in only one common stock. Today he obtains wide and well managed diversification of stock holdings by purchasing shares in good investment trusts." After the crash of October 1929 and the subsequent market collapse, opinion put part of the blame for the collapse on the rapid development of investment funds, but that was largely unfair. Their further development during the first subsequent bull market in the 1960s has shown that they fulfilled an essential role. In 1999 there were about eight thousand mutual funds in the United States.

2.3.3 Emergency procedures

The most primitive form of stabilization procedure consists in closing the Exchange. Wall Street was closed during the panic of 1873, from September 20–30 and during World War I, from August 1 to December 12, 1915. A similar procedure is to suspend trading for a stock when it turns out that the balance between buyers and sellers cannot be achieved. During the suspension, which can last from a few minutes to a few hours, new orders will be received and in this way one avoids price changes which are too abrupt. Similarly most futures markets impose upper limits on the scale of daily price fluctuations (Atkin 1989).

After the crash of October 1987 computers were blamed for having amplified the fall, but computers can also serve to implement stabilization rules. Thus a regulation was implemented in the computer system of the NYSE slowing down transactions as soon as the variation in the Dow Jones index exceeds a given amount (for instance 50 points in 1997) in a given time interval.

2.3.4 Stock options

We have already suggested that options can make investments in common stocks less risky by making hedging more efficient, but we have not yet examined if options have an overall stabilizing effect on price fluctuations. Without going here into a general discussion of this much debated question one can note the following points.

• By the end of December 2000 the share price for the e-commerce company Amazon.com was 15 dollars; at that time the strike prices of the stock option for July 2001 (i.e. six months ahead) ranged from 10 to 75 dollars. This large range reflected the high volatility of the stock in the last months of 2000, but obviously in a very conservative way. It should indeed be remembered that the price of this stock had increased from 1.3 dollars in July 1997 to 106 dollars in December 1999, before falling back to 15 dollars in December 2000. At that time, by merely extrapolating from former trends, the possibility that the stock could drop much lower than 10 dollars within the next six months could hardly be excluded (especially if one remembers that the earnings per share were still negative at that time). The fact that the range of strike prices did not extend beyond 10 dollars shows that stock option traders were rather cautious in their forecasts.

The same argument could be repeated for many other fast growth stocks. Thus for Yahoo, for instance, the price was 30 dollars by the end of December 2000 and the strike prices of options for July 2001 were restricted to the interval 15 to 100 dollars; since in 2000 the price had dropped from 216 dollars to 30 dollars a simple extrapolation of the recent trend would have suggested a possible price interval extending well below 15 dollars.

• The stabilization effect suggested by the above considerations is confirmed by a number of recent studies. (i) In the United States, Fedenia and Grammatikos (1992) found that the average bid–ask spreads on the NYSE declined by 15 percent after the introduction of stock options. (ii) For the Helsinki market, Sahlström (2000) found that for 13 optioned stocks the average standard deviation of raw returns decreased after the introduction of options; the decrease comprised between 18 and 31 percent according to the length of the estimation interval (from 120 to 250 days). (iii) It should be noted that on the American over-the-counter market (that is the automated transaction markets) the bid–ask spreads increased by 10 percent after the introduction of options (Fedenia and Grammatikos 1992). This suggests that the effect of options on the volatility depends on the organization of the market.

2.4 Sources of instability: the boomerang effect

Broadly speaking one can say that, depending on how they are used, most of the stabilizing factors can become sources of instability. In order to explain this point let us consider the case of options on stocks and indexes. As already mentioned most of these contracts expire on Friday at the end of the trading day. Now, observation shows that in the one or two hours before the bell there is often a fierce battle between bulls and bears. Depending on the outcome of the battle, the Dow Jones index can rally or fall sharply. Here are some examples.

• On March 20, 1987 (third Friday), in the last hour of the trading day, volume was so heavy that the tape was almost half an hour late at the close; eventually the Dow Jones lost 2 percent.
• The same scenario occurred on June 19, 1987 (third Friday), with a 1.5 percent fall of the Dow Jones.
• On October 16, 1987 (a third Friday) the Dow Jones dropped 8.6 percent. This was just before Black Monday when the Dow Jones lost about 20 percent.

As a result third Fridays came to be referred to as "Frightening Fridays" (Sobel 1987). The explanation of this instability was of course that option owners were trying to push the market in a direction which would permit them to exercise them profitably, a behavior that illustrates the boomerang effect mentioned in the title of this section. In the wake of the crash of October, the Exchange introduced rules which would limit that instability. Nevertheless the Friday effect persisted in subdued form, and the volume on such days remained well above daily average.

For instance, in mid 2000, Friday trade was about 40 percent above daily average both on the NYSE and on the NASDAQ. It is easy to check that such an effect did not exist before the introduction of the options: in September–October 1952 (two months chosen at random) Friday trade was on average 1.09 million shares, while the daily average for the rest of the week was 1.17 million. A more systematic study would be required of course, but at least qualitatively it seems safe to attribute the Friday effect to the introduction of stock and index options.

Many econometric studies have tried to examine whether the introduction and development of the futures market increased or decreased volatility. The question may depend on specific circumstances and is probably difficult to settle in general terms. Here is one example for which there is strong evidence that futures trading may amplify instability. On January 21, 1980, at the height of the silver price peak, the New York Commodity Exchange (Comex) suspended all futures and forward trading for ten days in order to limit speculation. After the suspension trade was resumed, but the number of contracts per investor was limited to 50 per month and their duration had to exceed six months (*Figaro*, February 2, 1980). Thus, it is clear that, in the eyes of the Exchange management, futures were considered as a major source of instability.

A number of other sources of instability are summarized in table 4.3. That in some critical circumstances specialists do not fulfill all their obligations is not pure speculation. As a matter of fact an investigation by the Securities and Exchange Commission after the crash of October 1987 has shown that in a general way the

Table 4.3. *Stabilizing forces versus sources of instability*

	Mechanism
Stabilizing forces	
Specialist system	That interface between market and investors cushions the effect of shocks
Investment funds	Averaging of shocks by diversification of stock holdings
New trading techniques	Makes hedging more efficient
Trading suspensions	Postponement of quotation whenever it would imply too large price changes
Computers	Implementation of automatic slowing down of trade whenever the price variation rate exceeds a critical level
Sources of instability	
Specialist system	Specialist may withdraw from the market on some critical occasions
Investment funds	Most funds have the same strategy
New trading techniques	Speculation on options and index futures affects the price of stocks ("Frightening Fridays")
Computers	Most trading softwares rest on the same assumptions

specialists did not assume the cushioning role they are expected to play; many of them had simply disconnected their phone and computer.

It can be (and has often been) argued that the great role played by investment funds is as a potential source of instability. The argument rests on the idea that a liquid market requires that at any moment in time there are both sellers and buyers, which in turn supposes a broad diversity of opinions among investors. Having attended the same business schools, fund managers may tend to react in the same way in a given situation, which will necessarily amplify market instabilities. This is no more than a plausible argument however. There were huge price fluctuations in stock markets even before the role of institutional investors became preponderant.

If the outbreak of a crash is conditioned by the organization of the exchanges, the unfolding of the crisis much depends upon the organization of the banking system. This is the question considered in the next section.

3 Organization of the banking system

In the nineteenth century, banking panics were defined as the suspension of convertibility of deposits into cash. In more recent times they are related to the inability of financial institutions to meet their obligations with their creditors. For instance a default on a corporate bond can have dramatic consequences for the bank through which it has been issued. In such circumstances the support it can get from other banks, or from the local (or federal) government, will be a factor of critical importance. In other words, the way banking panics unfold crucially depends upon the organization of the banking system. For instance, by providing a guarantee for depositors, the existence of a reliable lender of last resort tends to limit and soften the consequences of banking panics. As the question of banking panics will be touched on in the next chapter, it is of interest to discuss briefly some of the main issues in the organization of the banking system. Two different approaches are possible.

- One can describe the various institutions and regulations which structure the banking system and then try to *infer* how such a system is able to weather major crises.
- One can focus on how the banking system *actually* responded to various exogenous shocks. This is what can be called a black box approach in the sense that one does not try to understand in detail the internal mechanisms.

The difference between these approaches can be illustrated by drawing a parallel with the rating of a motor car. In the first approach one would open the hood, examine the engine and suspension, evaluate the aerodynamics of the bodywork, and then try to infer the performance of the car. The second approach would consist in placing this car with a number of others on the starting line of a racing track and observing the outcome of the race. The two approaches are clearly complementary but the first is far more difficult than the second. Even a careful investigation by experienced mechanics will probably fail to reveal which car is going to win the race. Why is it then that in economics the first approach is used much more often than the second? The reason is probably that the second implies a comparative analysis. In this section we try to use the second approach.

3.1 The United States

One can conduct a comparative study without leaving the United States for, as shown in table 4.4, there were several successive banking systems. Several major reforms were passed in the aftermath of severe crises. During the era of the National Banking System (1863–1914) the worst case in terms of failures of national banks was the panic of 1893 with a failure rate of 1.3 percent. During the Great Depression the percentage of national banks that failed was about 25 percent. In order to draw any conclusion as to the respective performances of the National Banking and Federal Reserve systems, one must find a way to gauge the severity of the exogenous shocks in each case. One possibility is to use downgrading ratios. Let us recall in that connection that corporate (or municipal) bonds are graded by rating companies, such as Moody's or Standard and Poor's, on a scale that ranges from Aaa (best quality) to C (bonds which are in default or have other marked shortcomings). The percentage of downgrades (or the ratio of downgrades to upgrades) can help in the estimation of the strain put on the banking system; comparing that figure with the ensuing failure rate may permit to gauge the robustness of the banking system. However, such an inquiry would lead us too far away from the topic of this book; it will be considered in a subsequent publication.

In the United States, not all states had the same banking system and this creates another possibility to carry out meaningful comparisons. Indiana and Ohio provide interesting cases, because besides the free banks they also had coinsuring banks. During the crises of 1854 and 1857 50 percent of the free banks failed, while none of the coinsuring banks failed. In general, a state-by-state investigation reveals that branch-banking systems tended to be less vulnerable (Calomiris and Gorton 1991).

3.2 Canada versus the United States

Unlike the United States, Canada allowed nationwide branching and relied on the coordination of a small number (40 in the nineteenth century falling to ten by 1929) of large banks to act as lenders of last resort. The average yearly failure rate of national banks during the period 1870–1909 was 0.1 percent in Canada as compared to 0.36 percent in the United States. Observation thus seems to suggest that the stronger the links between banks the better they can resist exogenous shocks. Needless to say those close links may also erode competition.

Table 4.4. *Organization of the banking system in the United States*

Period		Major crises
−1863	Branch banking laws in some states	
1863–1914	National Banking System	1878, 1893
1914–	Federal Reserve System	1929–1932
1933	Creation of the Federal Deposit Insurance Corporation (FDIC)	
1934	Creation of the Securities Exchange Commission (SEC)	
1991	FDIC Improvement Act	1990

Notes: Several reforms (1933, 1934, 1991) were passed in the wake of severe banking crises. To get an estimate of the gravity of the crises mentioned in the last column see the bank suspensions chart.

In the United States the attitude of the federal government has oscillated between two poles: on the one hand, to discourage banking combines, which would interfere with fair competition, and, on the other, to provide a guarantee at the national level in order to avoid widespread failure. But history seems to suggest that a pyramided system, such as the National Banking System (with several large New York City banks at the apex) or the Federal Reserve System (with the Fed at the apex), does not provide the same level of security as provided by strong inter-bank links. Thus the Fed was created in 1914 to ensure that illiquid banks would always be able to obtain enough cash to meet depositor demands, but the magnitude of bank failures during the Great Depression seems to suggest that the system did not meet the expectations of the legislator. In the aftermath of the Great Depression federal control was strengthened both for the banking system and the stock exchanges (see table 4.4).

So far the present chapter is mainly descriptive and qualitative. In the last section we present several time series which provide some quantitative landmarks.

4 Time series for stock prices and bankruptcies

4.1 Stock prices

Despite their well-established reputation for accuracy and completeness, German statistical yearbooks do not provide any information about share prices before 1885; similarly, share prices are to be found in Austrian statistical yearbooks only after 1877. Even in the excellent 'Historical Statistics of the United States' (1975), the statistics for common stocks' prices start only in 1871. In presenting charts of stocks' prices going back as for as possible our objective is to give a means to identify major speculative peaks and collapses. In terms of trends in prices there are marked differences between the three cases that we examine (figs. 4.3(a),(b),(c)).

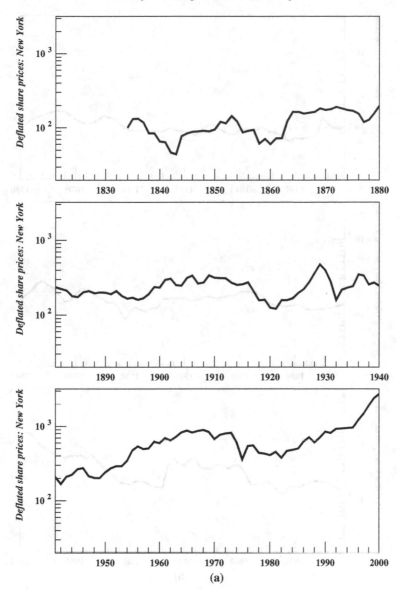

Fig. 4.3a. Share prices in New York 1834–2000
Notes: Vertical scale (logarithmic): deflated share index, 1834 = 100; in order to facilitate comparison the vertical scale is the same on two subsequent figures for London and Paris respectively. Until 1870 it is a railroad index, after that date it is the Standard and Poor's 500. Over the whole period the annual increase rate was 2.0 percent, but between 1940 and 2000 it was more than twice as fast, namely 4.4 percent as compared to 3.3 percent for the real gross national product. The chart shows that the panic of 1857 was only the last step in a stock market fall that had begun in 1853. The peak level of 1929 was surpassed only in 1956; the peak level of 1968 was surpassed in 1992.
Sources: 1834–1870 : Cole and Frickey (1928), 1870–1970: Historical Statistics of the United States (1975), 1971–1988: *Statistical Abstract of the United States* (various years), 1999–2000: http://finance.yahoo.com

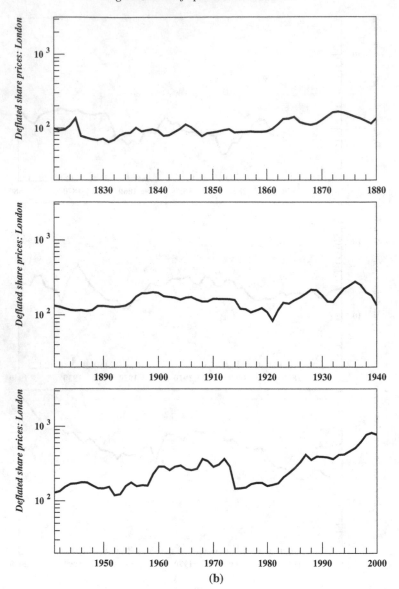

Fig. 4.3b. Share prices in London 1821–2000
Notes: Vertical scale (logarithmic): deflated share index, 1821 = 100. Over the whole period
the annual increase rate was 1.1 percent, which is 1.8 times slower than in the United States;
between 1940 and 2000 the increase rate was 2.9 percent (i.e. almost the same rate as for the
real gross national product). The peak in 1825 was entirely due to mining industries. The
chart shows that the crisis 1853–1857 in the United States had no impact on the London
equity market. Until 1918 the American and British series moved almost independently
one from another; subsequently their changes became fairly parallel but there is a greater
volatility of the New York market; this is particularly obvious during the 1921–1933 speculative episode.
Sources: 1821–1980 Mitchell (1988), 1981–2000: http://finance.yahoo.com

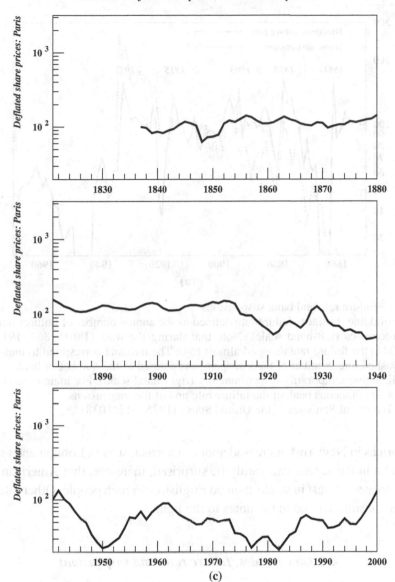

Fig. 4.3c. Share prices in Paris 1834–2000
Notes: Vertical scale (logarithmic): deflated share index, 1834 = 100. Over the whole period the annual increase rate was 0.16 percent, much slower than in the United States or in Britain; in the 1940–2000 period the increase rate was 1.6 percent. The crisis of 1853–1857 in the United States had almost no impact on the Paris equity market. In the nineteenth century the major speculative episode culminated in 1881; in January 1882 the failure of the "Union Générale" marked the end of the bull market. Note that the peak of 1929 was followed by a decade of declining prices, whereas in the United States there was a peak in 1936. It can be observed that there was a strong and rather surprising increase in stock prices in 1941 and 1942: the (deflated) index rose from 307 in 1940 to 571 and 762 respectively.
Sources: 1837–1855: Reznikow (1990), 1856–1960: *Annuaire Statistique de la France* (*Rétrospectif* 1966), 1961–1988: *Main Economic Indicators. Historical Statistics*, OECD (1989), 1988–1990: *Annuaire Statistique de la France* (1992), 1990–2000: http://finance. yahoo.com

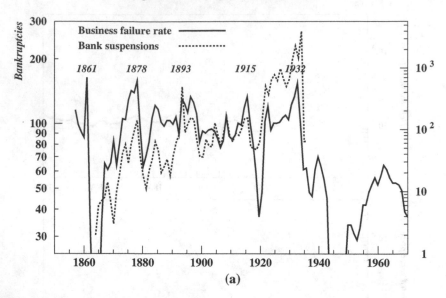

Fig. 4.4a. Failure rate and bank suspensions
Notes: Solid line: business failure rate defined as the annual number of failures per 10,000 listed enterprises (left-hand scale). Note that during the wars (1861–1864, 1917–1918, 1941–1944) the failure rate dropped almost to 0. The maxima correspond to major recessions. Dashed line: number of bank suspensions defined as the number of banks closed to the public either temporarily or permanently (right-hand scale). For main recession years there is a simultaneous peak of the failure rate and of the suspensions.
Source: Historical Statistics of the United States (1975: 912, 1038).

Stock prices in New York increased about twice as fast as in London and ten times faster than in Paris. One can hardly be surprised, therefore, that American people show a greater interest in stocks than do English or French people. Other interesting features are summarized in the notes to the figures.

4.2 Downgrades, failure rate, and suspensions

A panic can be defined quantitatively by a peak in the number of bank suspensions; a recession can be defined by a peak in the failure rate (fig. 4.4(a)). The downgrade variable is defined as the number of bond issuers whose quality was degraded from A to Baa. Let us recall in this respect that for good quality bonds (also called investment grade bonds) there are four classes: Aaa (best quality), Aa, A, Baa. Under Baa begin the so-called speculative grades (also called junk bond quality grades): Ba, B, Caa, Ca, C. The transitions from A to Baa and from Baa to Ba are particularly crucial, because Baa is the last grade before junk status. For instance, in December 2000 Xerox was downgraded from Baa2 (the Baa class is subdivided

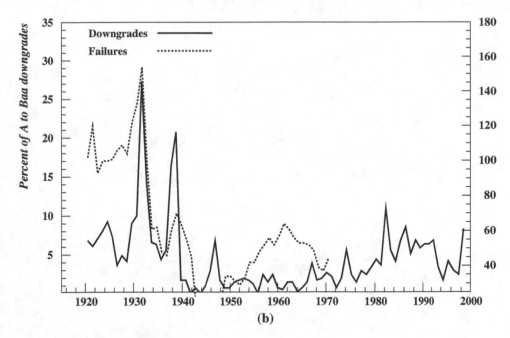

Fig. 4.4b. Downgrades and failures
Notes: Solid line: annual percentage of bond issuers downgraded by Moody's from A to Baa (left-hand scale). Dashed line: failure rate (right-hand scale). The correlation of the two variables is 0.62.
Source: Downgrades: Rating methodology. Moody's Investors Service (August 1999); failure rate: see previous figure.

into Baa1, Baa2, and Baa3) to Ba; as a result, if Xerox wanted to issue bonds it would have to pay a much higher interest rate than a Aaa issuer.

For a given company a failure is usually preceded by one or several downgrades, and one would, therefore, expect downgrades to provide an early warning of an impeding recession. However, fig. 4.4(b) shows that this is hardly the case. The downgrade and failure curves are almost synchronous, and, if there is a delay, it is certainly smaller than one year.

Part III

Regularities in speculative episodes

5

Collective behavior of investors

As a businessman established in California, Chuck Hazel is not scared by talk of a recession or predictions of soaring oil prices. He isn't alone. Prospects have seldom looked so good for entrepreneurs who, like him, head tiny high-risk companies who make innovative products. A key reason is that, despite high interest rates and fears of a business slowdown, money for investing in such risky ventures is plentiful. So-called hot areas are the entire data-processing and communication markets, medical care with new diagnostic devices and synthetic materials. Much of the seed money for these young companies and for individuals with a dream for a new product comes from professional venture capital firms that specialize in financing start-ups, expansions, and repurchases. Arrangements vary widely, but often a venture capitalist will provide, say, half a million dollars in return for a share of the privately held stock.

(*US News and World Report, December 24, 1979, adapted*)

This account describes fairly well the high-tech boom that led to the Internet revolution of the 1990s, but in fact it was written in 1979. As a matter of fact the high-tech boom of the late 1970s comprised many of the ingredients that would become famous in the 1990s: the dreams of individual innovators in the sector of data processing, the high-growth companies established in California, the crucial role played by venture capital, the bonuses payed to the staff in the form of stock options, the amazing amounts of money raised by initial public offerings (i.e. sale of the company's stock to the public). High-tech booms are recurrent events in the economic history of the United States and they are a good illustration of the purpose of the present chapter which is to show that economic events (like most other historical events) do not occur out of a clear blue sky, but are preceded by forerunners and followed by replicas. In other words, instead of studying single events we will define and analyze *clusters of similar events*.

For each of the phenomena considered in the three following chapters the steps will be the same. First, we focus our attention on a specific phenomenon. Second,

83

we define and analyze a cluster of similar events which feature that phenomenon. Depending on the data available, we will be able to carry out a quantitative comparison or we will have to content ourselves with a qualitative analysis; in either case our objective is to find meaningful regularities. For some readers the fact that at this point we refrain from proposing any model may perhaps seem disappointing. However, stock market crashes (or, more generally, the crashes of speculative markets) are not a homogeneous and well-defined class of phenomena. Some are triggered by problems on the bond market, e.g. the 20 percent drop of American markets in mid 1990 following the junk bond crisis; some are provoked by brutal shifts in exchange rates, e.g. the 48 percent fall of the Mexican equity market following the devaluation of the peso in December 1994 or the 1997 crisis in South-East Asia; some are caused by rising interest rates, e.g. the 23 percent fall that occurred in American markets between November 1980 and July 1982. Our ultimate objective is to investigate *underlying mechanisms*, i.e. what Harvard sociologist Stanley Lieberson (2000) calls internal mechanisms, but before that can be done fruitfully we must first disentangle fundamental phenomena from those which are merely accessory and circumstantial.

1 High-tech booms

We begin our investigation of high-tech booms by considering the beginnings of the automobile industry in the early 1900s.

1.1 The high-tech boom of the automobile industry

At first sight nothing could seem more different from Internet start-up companies than the automotive industry with its huge companies, its dozing technological development, and its fierce international competition. The situation was very different at the beginning of the twentieth century however. In 1923 there were still only 18 million motor vehicles in the world and 83 percent of them were in the United States. More precisely there were 14 motor vehicles per 100 inhabitants in the United States against 1.5 in the United Kingdom or 0.25 in Germany (*The World Almanac and Book of Facts 1925*). But the main difference with the present day is that it was possible to start an automobile company with very little capital. As a matter of fact, at its beginnings the automobile industry had several of the characteristics that people came to associate with the "new economy" of the 1990s: very rapid growth rates, localization in a fairly limited region, insensitivity to changes in interest rates, and high stock valuations. Let us consider these points in more detail; in so doing we draw heavily on a fascinating and very stimulating book by Doerflinger and Rivkin (1987).

1 To start a company an entrepreneur had to build a modest shop costing perhaps 5,000 dollars, hire a dozen workmen each at 1.25 dollars per day, and set them to work bolting together the basic parts of the vehicles that were produced by various subcontractors. The Ford Company was started in the early 1900s with a capital of just 28,000 dollars, and within 15 months had produced a profit amounting to ten times that sum. Not all companies were so successful however; while 485 companies entered the market between 1900 and 1908, eight years later there were only 243 survivors.

2 Between 1905 and 1908 the annual growth rate of General Motors was about 85 percent. Overall, between 1906 and 1923 the sales of motor vehicles increased at an average rate of 24 percent per year.

3 In the same way as the data-processing industry of the 1980s and 1990s was concentrated in the Silicon Valley, the automotive industry was located in a fairly small area around Lake Erie in Michigan and Ohio. In less than a decade a gigantic new industry sprang to life in the cities of Akron, Detroit, Flint, and Toledo.

4 Between 1910 and 1920 the price of General Motors' shares was multiplied by 12 (*Common Stock Price Histories*, 1988). Similarly during an earlier high-tech boom the price of American Bell stock was multiplied by 20 in a few months (between March and September 1879). Needless to say such rapid increases resulted in stratospheric price to earnings ratios.

The previous observations point out that the Internet boom followed the standard pattern set by previous high-tech booms that were of the same kind. However, it must be emphasized that not all high-tech booms followed the same pattern; thus, the establishment of railroad networks necessitated long-term mobilization of huge amounts of capital before the first earnings would be collected. Similarly in the biotech industry there are very long delays between a scientific discovery and the moment when a drug can be marketed.

1.2 The phase of "natural selection"

The history of past high-tech booms also gives us a better insight into how they end. When the ebullience of the market abates a selection process sets in which can be beneficial, but which can in some cases go too far. We have already mentioned that in the 1900s the number of automotive companies dropped by 50 percent to about 250; needless to say that number was further reduced in later decades. Similar data can be mentioned for the computer high-tech boom of the late 1960s: from the 46 companies which went public about 37 percent no longer existed 12 years later. In the same vein one can remember that a number of the success stories of the 1970s and early 1980s, such as Atari, Commodore, Victor Technologies, or Fortune Systems, had a fairly brief existence.

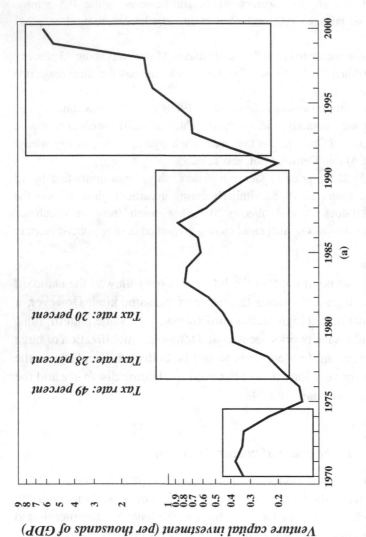

Fig. 5.1a. Three high-tech booms driven by venture capital investments

Notes: The figure identifies three peaks of increasing amplitude during the late twentieth century. All three were largely connected with information technology; broadly speaking, the first was based on new silicon chips, the second on the diffusion of personal computers, the third on the Internet revolution. The second peak is all the more remarkable because 1979–1982 was rather a recession period for the US economy: GDP growth was negative in the third quarter of 1980 and during a large part of 1982. Note that the amplitudes (defined as the ratio of peak value to initial value) of the peaks increase from 2 to 5 and 30.

Sources: Geoffron (1990), Beacon Management (Internet 2001).

Even successful companies can become easy prey for banks and financial institutions. A case in point is General Motors. In the summer of 1910 the combination of over-expansion and the acquisition of Cadillac created a liquidity squeeze. A syndicate headed by New York banks floated a 15 million dollar bond issue of five years. In return it insisted of getting managerial control of the company during the life of the bond, an episode which bore a strong resemblance to the bankers takeover of American Bell in 1877. Doerflinger and Rivkin (1987) convincingly argue that control by financial institutions usually led to a sterilization of the company's capacity for innovation.

In a sense the selection process which follows high-tech booms is not without some definite benefits for the industry, as it trims off non-viable companies and makes survivors become more patient, less enamored of the stock market, and more intent on building over the long term, a shift which was suggestively summarized in the following terms by a manager (Doerflinger and Rivkin 1987): "Today there is the requirement that you be a hell of a lot more intrinsically interested in the companies than you were before, because you may be in them for a hell of a lot longer." Needless to say, the duration and severity of the recessions that follow high-tech booms are largely determined by national and international macroeconomic factors, which explains why in some cases the process may go too far and put out of business some perfectly sound companies.

So far we have only briefly mentioned the important question of how high-tech booms are financed. The development of the US venture capital industry after World War II was a major step in that respect. In the following paragraphs we briefly describe the changes that it brought about; one of the most important was a closer connection between high-tech companies and the stock market with the correlative multiplication of initial public offerings.

1.3 High-tech booms backed by venture capital

The creation of the venture capital industry was made possible by the observation that the early stages in the creation and development of a company could be standardized, and as always standardization permitted a great leap in social productivity.[1] Instead of leaving each entrepreneur to find out by himself how to obtain financing and find offices and staff, the so-called incubator companies offered office space, accounting services, and financing. The objective was to bring a start-up company close to the stage of initial public offering (IPO) in a shorter time interval. Fig. 5.1(a) shows that there were three major waves of venture capital investment. Tax policy

---

[1] The very same idea was used with great success in various business sectors; for instance through its uniform worldwide organization a company like McDonald's is able to realize tremendous economies of scale at all levels (planning, equipment, food, management, and so on).

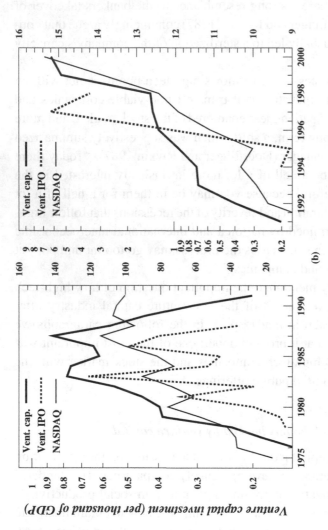

Fig. 5.1b. Two high-tech booms

Notes: The charts compare three factors: (i) the annual venture capital investment, (ii) the amount of venture capital which led to initial public offerings (IPOs), and (iii) the changes in the deflated NASDAQ Composite index. In both charts the left-hand scale refers to venture capital investment and the right-hand scale to IPOs of venture-capital-supported companies (number of companies for the 1975–1990 chart, amount of capital for the 1992–2000 chart). It can be noted that the fall of the NASDAQ index in years 1986–1990 which were marked by a contraction of venture capital investments was rather modest: it slightly decreased between 1986 and 1990 and resumed its ascension in 1992.

Sources: Venture capital: see fig. 5.1a; IPOs: Geoffron (1990), Venture Coach (Internet 2001), NASDAQ: *Statistical Abstract of the United States* (various years).

and the possibility for pension funds to invest in venture capital were obvious amplification factors which explain the magnitude of the investment in the 1990s.

The similarity between two high-tech booms of the late twentieth century is documented in fig. 5.1(b). Although venture capital represented by 2000 only about 4 percent of total business investment this seed money played a great role. Only a small fraction of the firms financed (partially or totally) by venture capital went public, but through the money they generated these IPOs were a major driving force in the whole process.

When a boom ends there begins for investors what is usually called a flight to safety; this is an essential component of any speculative bubble. Whether the burst of the bubble will lead to a short-lived or protracted recession to some extent depends on the investors' behavior in this critical phase. This is the subject of the next section.

2 Flight to safety

A flight to safety occurs when investors face the risk of heavy losses either because of a sudden fall in the price of an asset or a security or because of a banking panic. First of all let us illustrate this kind of event with the panic of October 13, 1857, as described by the correspondent of *The Times* (it was published on October 27, p. 9). In this case the panic can be observed in the form of a street demonstration; in twentieth-century panics, underlying collective phenomena may have been similar but no longer gave rise to open demonstrations, and, as a result, they cannot be observed as easily.

The first run yesterday was made upon the smaller banks outside of Wall Street, that afford circulation for the artisans, shop-keepers, hotel-keepers. These institutions were naturally in a less strong position than the banks doing business with the mercantile classes. Up to 1 o'clock everything was quiet. There was a steady payment of specie over the counters to depositors, but nothing indicating a general alarm. Then, almost in an instant, the street was crowded and a run begun upon the American Exchange Bank, the weakest of the large institutions. There was a crowd of some hundreds (or thousands rather) gathered in front, and a long line of bill-holders formed a queue. Mr. David Lemitt mounted a step, and treated the crowd to the universal Anglo-Saxon panacea, a speech. The crowd partially dispersed but the bill holders kept up the run. From the American Exchange Bank the attack shifted to two or three banks further up the street. The Bank of America and the Mechanics Bank were particularly selected. As [the people] came out with pockets bulging out with gold some looked happy, more looked uneasy and foolish.

Table 5.1. *Flight to safety*

Time period	Item	Safe shelter
–1860	Grain	Hoarding grain
1700–1850	Currency	Gold and coins
Inflation time	Currency	Tangible assets
Stock crashes	Stocks	Blue chips, value stocks
Stock collapses	Stocks	Treasury notes

Notes: As illustrations of the shift to government bonds one can mention September 1998 and January 2001; in both cases the price of two-year Treasury notes experienced substantial increases. The shift from growth stocks (bought for capital gains) to value stocks (bought for dividends) can be illustrated by the contrast between 1999 and 2000: in the first year the Russel 3000 Growth Index outperformed the Russel 3000 Value Index by 27 percent, while in 2000 the Value Index outperformed the Growth Index by 30 percent.
Source: Webfund Report, January 4, 2001.

From Wall Street the rush extended to Pine and Nassau [streets] and the large Broadway banks; before 3 o'clock 18 banks fell, with a united line of loan of 21 million dollars.

This account is interesting in several respects.

• Safety was represented by cash, but the process is much more general. A break of confidence regarding a given security leads to a flight to safe havens. A number of cases are summarized in table 5.1.

• The panic of October 1857 has been described in several historical accounts of financial crises (e.g. Wirth 1883), but it has often been misinterpreted as the beginning of a stock market crash similar to 1929. In truth, the October panic was a purely monetary panic, which marked the *end* of the stock market fall. By October 17 the *New York Herald* already mentions a strong stock market; note that for some stocks (e.g. New York Central, New York and Erie, and Reding) the upturn did in fact occur before the panic. On the chart of stock prices given in the previous chapter it can be seen that after the peak, which occurred in early 1853, there was a long slide until March 1857, during which the market lost 36 percent; then, between March and October, there was a sharp fall with a further loss of 50 percent, and, between October 1857 and January 1858, share prices rebounded and gained 60 percent. Incidentally it can be noted that the 1929–1933 crisis was marked by a similar succession of events, in the sense that the height of the banking crisis occurred two or three years after the stock market crash.

The flight from a hazardous position to a safe shelter is a natural behavior; at an individual level it is even fairly banal. But we are interested in collective rather than individual behavior, and at that level the phenomenon is far from being banal.

All of a sudden a large group of people, who in normal times are only loosely connected, start to behave in a synchronized way. The more interconnected the people become the stronger is the movement. Once begun such panics proceed unrelentingly. In the following subsections it will be shown that similar mechanisms can be observed in different contexts and for different time periods. It will be notable progress if, instead of looking at these episodes as so many different cases, one can analyze them as different facets of a single phenomenon.

2.1 Grain panics

In Western Europe before 1850 social disturbances connected with grain shortages were common. Economic historians and sociologists have given accurate descriptions of such episodes; see for instance the seminal studies by Abel (1966), Meuvret (1971), Tilly (1983, 1992), and Miller (1999). At first sight it might seem somewhat farfetched to make a parallel between subsistence crises and the flight to safety analyzed in the rest of this section. This is justified by the following considerations: (i) Most grain crises were not characterized by widespread grain shortage; grain remained available, albeit at an inflated price. (ii) Usually grain crises lasted for five to eight years, and it is, therefore, difficult to attribute them to one or even two bad harvests; reduced yield was rather a triggering factor which started a speculative spiral. As belief in an imminent price increase spread, people began to buy huge amounts of grain in anticipation. This process is well documented by contemporary writers, see, for instance, Biollay (1885), Modeste (1862), or Martin (1908). It is hard to draw a frontier between precautionary purchases and speculation; clearly for people who held large inventories it was natural to sell at the most profitable price.

Why should one care about grain crises in a study of speculation? What makes them important is the fact that they lasted for several centuries. Anxiety about scarcity repeatedly actuated collective movements, which left a lasting mark in the collective historical memory. Grain panics were the social models for subsequent banking panics; the people of October 1857 who stormed one bank after another had a strong resemblance with those who stormed grain reserves or bakeries. Moreover, throughout the nineteenth century, grain panics were synchronized with specie crises. This is the subject of the next subsection.

2.2 Nineteenth-century banking panics

Until 1972 the dollar was convertible into gold on the basis of 888 grammes for one dollar; the Swiss franc clung to gold convertibility until the end of the twentieth century. Seen in a historical perspective a non-convertible currency is a very recent innovation. It must be remembered that the first experiments with paper currency (the Law episode in France around 1720, the experiences during the American and French Revolutions) ended in disaster. Therefore it should be no surprise that, throughout the nineteenth century, the public was concerned about the safety of savings held in bank accounts.

In the excerpt at the beginning of this section one reads that "the people came out of the banks with pockets bulging with gold"; that may well be a stylistic exaggeration however. After all in 1857 bi-metallism was still in force (it would be abolished only in 1873) and the pockets should, therefore, have been bulging with silver as well as gold. More to the point, one may wonder whether the lack of confidence really extended to the greenbacks themselves or if it rather remained confined to banking institutions. For the panics that occurred in the late nineteenth century it is certainly the later interpretation which holds. This is further confirmed by the fact that for some banks the withdrawal took a far less dramatic form. As documented by Kelly and O'Grada (1999), between September 28, and October 13, 1857 over 500 depositors of the Emigrant and Saving Bank in New York closed their accounts, but this represented only 9 percent of customers and did not imperil the bank.

One might think that currency panics definitely belong to a bygone era. That may be the case in industrialized countries but, as suggested by the following example, it is certainly not true of developing countries. As a matter of fact, in June 1999 there was an outburst of bankruptcies in Uganda; within two days several thousand people closed their accounts and half a million dollars was withdrawn in cash.

2.3 Relationship with grain crisis

Clément Juglar (1862) convincingly showed that, at least for the first half of the nineteenth century, financial panics used to follow grain price peaks by one year (table 5.2). The mechanism could be decomposed into the following steps: (i) The high grain prices drained currency from the market because they transferred large amounts of cash into the pockets of wealthy people (big farmers, wholesale traders, and various speculators), who did not immediately recycle them into the monetary circuit. (ii) This depleted the cash reserve of the central bank. (iii) To counter the outflow of cash the discount rate was temporarily increased. The fact that a large share of the people's income went into grain purchases coupled with the higher discount rate led to an economic slowdown. Usually, however, the recession was short, typically one or at most two years.

Note that banking panics were even shorter; usually they did not exceed two or three months, and their analysis consequently requires high-frequency statistics which are not easy to find. For instance, between January and November 1825 the cash reserve of the Bank of England decreased at a fairly constant monthly rate

Table 5.2. *Relationship between grain price peaks and currency crises France, 1800–1860*

Price peak, year (n)	1803	1812	1817	1829	1839	1847	1856
Price at peak (franc): p_n	23.38	34.41	36.16	22.59	22.14	29.01	30.75
Start of currency crisis (m)	1804	1813	1818	1830	1839	1847	1857
Δ(cash reserve) $([m] - [m - 1])/[m - 1]$	−10%	−58%	6%	−35%	−14%	−22%	−21%
Bills discounted $([m + 1] - [m])/[m]$	−60%	−87%	−37%	−64%	−11%	−48%	−30%

Notes: As a matter of comparison the average price of wheat was about 15 francs per hectoliter. The scenario suggested by these figures comprises the following step: (i) High grain prices disrupt the channels of money flows by concentrating unusually large amounts of money into the hands of big producers and traders. (ii) One year later this results in a diminution of the cash reserve of the central bank (the figures given above are yearly minima); the central bank reacts by increasing the discount rate. (iii) The fall in consumption coupled with the higher discount rate brings about a recession as shown by the dramatic fall in the amount of discounted bills (in the absence of GDP data the latter reflect the business slowdown).
Sources: Juglar (1862), Labrousse, Ramano, and Dreyfus (1970), *Annuaire Statistique de la France* (1887).

of 10 percent; then between November and December it suddenly dropped by 60 percent (Juglar 1862).

2.4 *"Deliver us from inflation"*

The title of this paragraph is taken from a book published in Canada in 1957 (McConnel 1957). Although inflation was low (slightly over 2 percent) throughout the 1950s, both in Canada and in the United States, this title aptly suggests that inflation is a major fear of investors. As a matter of fact the reactions of investors to double-digit inflation rates are interesting to study in a comparative perspective, for they are largely independent of the specific business situation. The most common reaction is to invest into tangible assets that can provide a hedge against inflation: gold, silver, real estate, rare coins, and other collectibles. In order to substantiate this assertion one has first to identify time intervals and countries where inflation assumed two-digit rates; a second step is to procure adequate price series for tangible assets. Fig. 5.2(a) summarizes inflation figures for two cases characterized by substantial inflation levels. The first case is France between 1930 and 1947. The first years of the 1930s were marked by deflation, but inflation rose to and remained over 10 percent in the second half of the decade and assumed even higher levels during World War II. The second example is Britain in the 1970s. These were inflationary years in almost all industrialized countries, but the inflation levels were particularly high in Britain. A similar example was the United States for the same period. In the following subsections the flight to tangible assets will be illustrated by the examples of postage stamps, antiquarian books, and precious metals.

2.4.1 Postage stamps

The increase in the price of postage stamps in France during World War II was already documented in chapter 3 for two specific stamps. In a more general way the deflated price index of all stamps jumped from 84 in 1939 to 284 in 1944 (Feuilloley 1996); this corresponds to an annual increase rate of 24 percent. In order to study the price increase in Britain we computed a (deflated) geometric average for a sample of nine stamps. It climbed from an index value of 100 in 1970 to 274 in 1980; this corresponds to an annual increase rate of 9 percent.

2.4.2 Antiquarian books

Between 1934 and 1945 the deflated price of books written by the French author Victor Hugo increased by 38 percent, which corresponds to an average annual increase rate of 3.2 percent.

For Britain, instead of considering the books written by a given author, we studied the price movements of random samples of 150 books. Needless to say, the samples for successive years depend on the books offered for sale at the auctions and may comprise very different items. Thus, the sample price average may possibly be biased by the inclusion of some very expensive books. In order to avoid this problem the samples were decomposed into four quartiles, and in this way only the fourth quartile (i.e. the 25 percent highest prices) can possibly be biased by the inclusion of very expensive books. For the first quartile (the quartile of the cheapest books) the average price increase between 1971 and 1979 was 32 percent; for the second and third quartile it was 15 and 16 percent respectively; not surprisingly for the fourth quartile the movements were completely erratic as the average mainly reflected the fact that some very expensive books were (or were not) included in the sample.

Fig. 5.2(b) illustrates the same phenomenon for the United States. Using the same methodology as for Britain we represented the movement of average prices for three quartiles; the fourth quartile (highest prices) again shows fairly erratic movements and was left out. Between 1976 and 1980 real prices increased by 80 percent, which corresponds to an annual rate of 17 percent.

As far as inflation is concerned the most spectacular example in industrialized countries was Germany between 1913 and 1923; the price increase began

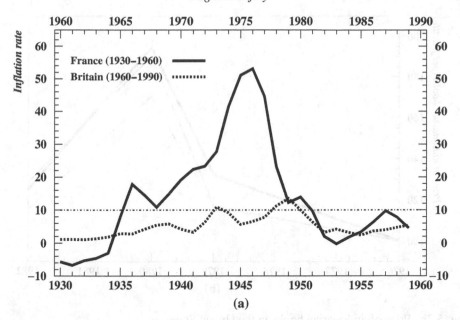

Fig. 5.2a. Two inflation episodes
Notes: A three-year moving average was performed. As a rule of thumb one can retain the "double-digit criterion" which says that (at least in industrialized countries) the behavior of investors begins to be seriously affected by inflation whenever the inflation rate exceeds 10 percent. The inflation episode of the 1970s was common to all industrialized countries but inflation rates were particularly high in Britain.
Source: Liesner (1989).

progressively (between 1913 and 1921 the average rate was about 38 percent) and ended (between mid 1922 and November 1923) in a staggering hyperinflation. Did the deflated price of antiquarian books increase during that period? During the first phase of the inflation it did not; once corrected for inflation the average prices (in the three first quartiles) *decreased* from a level of 100 in 1913 to about 12 in 1921. The obvious reason is probably that during the war there were only a few auction sales and thus book prices were not able to catch up with inflation. The investigation of the second phase between mid 1922 and November 1923 would require a high-frequency analysis: at a monthly level in 1922 and at weekly (or even daily) level in the fall of 1923; that study will be postponed to a subsequent publication.

2.4.3 Precious metals

Between 1977 and 1980 there was a rapid increase in the prices of cobalt, diamonds, gold, palladium, platinum, and silver. The average annual increase rate of real prices

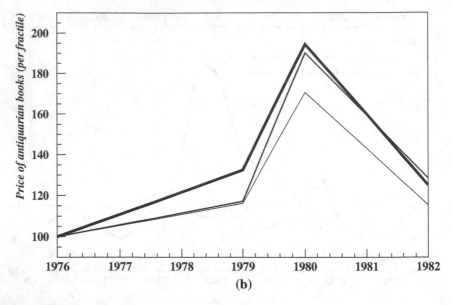

Fig. 5.2b. Price of antiquarian books in the United States
Notes: Book samples have been divided into four quartiles: the average price of the 25 percent lowest prices corresponds to the thin line; the two following quartiles are represented by thicker lines; the highest quartile has not been represented because it is highly dependent on the prices of a small number of very expensive books.
Source: American Book-Prices Current (various years).

was of the order of 43 percent; the interested reader can find more detailed figures in Roehner (2001a, 69).

2.4.4 Conclusion

The previous results are summarized in table 5.3. Precious metals seem to be the most popular hedge against inflation, followed by postage stamps and antiquarian books. This observation seems plausible for it is fairly easy to buy a contract for silver on a futures market; in contrast, postage stamps and books are non-standardized items, whose price can be soundly estimated only by an expert.

In order to sharpen our conclusions one would need a larger sample of cases than those mentioned in table 5.3. For instance, we did not mention the Civil War period (1861–1865) in the United States, which saw a 108 per cent jump in retail prices, because no stamp or book prices were available. In a general way there are only few inflationary episodes in industrialized countries, which means that we are confronted by what sociologists call a "small N" situation; this is a problem that we will encounter repeatedly in following chapters.

Table 5.3. *Real price increases for tangible assets in times of double-digit inflation rates*

	Average annual inflation rate	Precious metals	Postage stamps	Antiquarian books
France (1939–1948)	32%		24%	3.2%
Britain (1972–1981)	14%	43%	9%	5.2
United States (1972–1981)	8.8%	43%		17%
Germany (1913–1921)	38%			−88%

Notes: The price data refer to real (i.e. deflated) prices. With the conspicuous exception of Germany (which may be due to war circumstances or to the fact that the price increase was too fast: 79 percent annually between 1919 and 1921) investors tried to find a hedge against inflation by buying tangible assets. The price increase for precious metals corresponds to the international market. Although more and more internationally integrated, the markets for stamps and antiquarian books were in the 1970s still mainly national.
Sources: Inflation rates: Liesner (1989); precious metals: *La Vie Française* (January 14, 1980); stamps: Massacrier (1978); books: *American Book-Prices Current*, Book auction records: a prices and annotated annual record of international book auctions, *Jahrbuch der Bücherpreise* (Otto Harrassowitz).

2.5 Flight to quality in equity markets

It is a common saying that in times of uncertainty in the equity market investors tend to fall back on (so-called) blue chip stocks (such as General Electric or IBM). It can be of interest to examine to what extent the present rule is corroborated by observation.

We first examine the behavior of investors during three pre-World War II crashes, namely 1907, 1929, and 1937. For this purpose we compare the course of prices for different kinds of companies in the weeks following the crash. By using the *Dow Jones Averages* edited by M. Farrel (1972) it is possible to observe on a daily basis the three Dow Jones indexes: industrials, railroads, and utilities (the last one only after January 1929). The notion of a "blue chip" company is of course fairly subjective, however for a company to have been in existence for several decades is probably one of the characteristics which can reassure investors. In the early twentieth century, railroad companies were almost the only companies which could claim to have been in existence for more than three decades (remember that before 1870 railroad companies constituted the backbone of the NYSE). In other words, it is not unreasonable to posit that railroad companies were at that time the senior companies, while the industrial and utility indexes mostly represented more recently introduced companies. Fig. 5.3(a) shows that in the months following the crash of March 1907 railroad companies behaved rather better than industrials: between

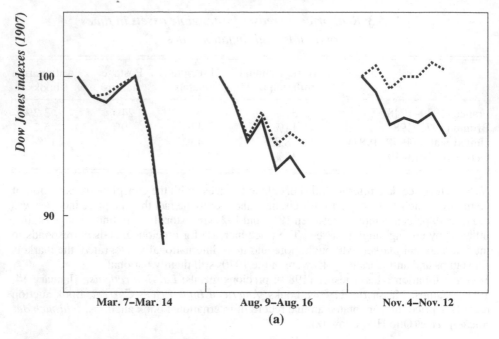

Fig. 5.3a. Flight to security during the market collapse of 1907
Notes: Solid line: Dow Jones industrials; dashed line: Dow Jones railroads. The market collapse began in November 1906 and the lowest point was reached one year later; the total price drop was 38 percent. The three weeks considered in the chart concentrated a total fall of 24 percent. Except in the first week, investors considered the railroad companies as a safe shelter in times of uncertainty.
Source: Farrel (1972).

August 9 and 16 the industrials dropped by 8 percent, while the railroads lost only 5 percent; between November 4 and 12 the industrials fell by 4 percent, while the railroads gained 0.6 percent. However, for an unknown reason, the preference for railroads is *not* observed in the first week after the crash.

The same analysis is carried out in fig. 5.3(b) for the crash of October–November 1929. This time the flight to railroad companies can be observed unambiguously even during the first week. In each of the four weeks the railroads perform notably better than the two other indexes. This observation is all the more striking if we remember that the medium-term performance of railroad companies in the two or three years after the crash was markedly below average. In other words the behavior of investors documented in fig. 5.3(b) did not reflect expectations based on fundamentals but rather a spontaneous flight reaction.

Fig. 5.3(c) tells a completely different story. The railroads experience a more pronounced fall than the two other indexes. In 1937 the railroads were no longer considered as a safe haven, probably because of their poor overall performance in the early 1930s.

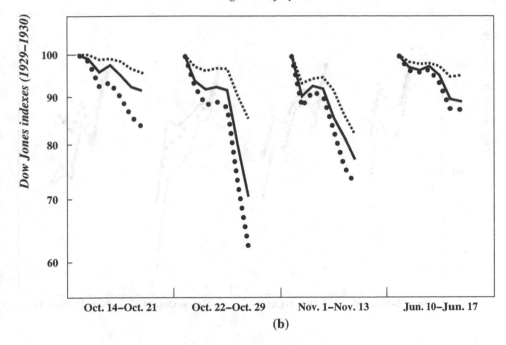

Fig. 5.3b. Flight to security during the crash of 1929
Notes: Solid line: Dow Jones industrials; dashed line: Dow Jones railroads; dotted line: Dow Jones utilities. Investors deserted the utility companies which had been the high-growth companies of the decade and sought refuge in well-established (albeit outdated) railroad companies.
Source: Farrel (1972).

The second set of evidence that we consider concerns the period following the NASDAQ crash of March 2000. Fig. 5.3(d) displays a typical episode. On April 3, 2000 the NASDAQ Composite index dropped by 6.7 percent, while at the same time the Dow Jones Industrials gained 2.8 percent. But on the following day the fall of the NASDAQ index accelerated and reached 14 percent in the morning of Tuesday. Under such circumstances one observes that the flight to safety is replaced by a joint fall (note that the NYSE fell much less: 5.3 percent against 13.9 percent). This positive correlation may be due to what is called the "margin call effect." As the value of their portfolio falls investors who bought shares on borrowed money will have to add some cash because the guaranty represented by the shares becomes insufficient to cover the loan. If the total amount of loans extended to buy shares (the so-called margin of debt) is large this can lead to a self-reinforcing phenomenon. In 1929 the margin of debt represented about 10 percent of the capitalization of the NYSE; in the spring of 2000 it represented only 2 percent. A succession of events, such as the one described in fig. 5.3(d) (i.e. a flight to safety followed by a joint fall), occurred repeatedly

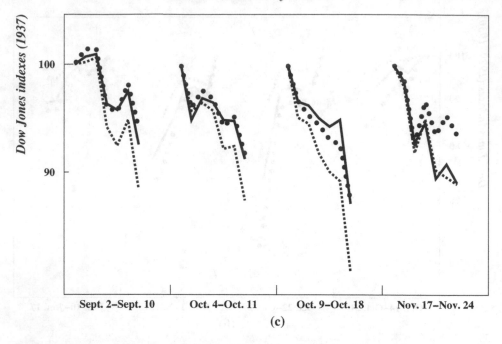

Fig. 5.3c. In 1937 railroad companies were no longer considered as a safe shelter
Notes: Solid line: Dow Jones industrials; dashed line: Dow Jones railroads; dotted line: Dow Jones utilities. In marked contrast to what happened during the crash of 1929, prices of railroad stocks experienced the worst fall. This may be attributed to the fact that their performances in the early 1930s had been disappointing in terms of stock valuation.
Source: Farrel (1972).

in the wake of the March 2000 crash; for instance on April 13–14, May 2–3, May 8–9.

Other "flight to quality" episodes are summarized in table 5.4. The blue chips of the Dow Jones industrials offered a hedge against the decline of the NASDAQ, but the utilities index rose even faster than the industrials index. Having been neglected by investors for a long time, utilities were undervalued and provided a natural hedge against an overvalued NASDAQ.

For experienced investors it may be possible to hedge against a fall in one sector by shifting their assets to another; that possibility is documented for the year 2000 in table 5.5. However it must be noted that 2000 was a fairly exceptional year in that respect.

The previous observations may suggest that investors react very rapidly (within one day) to market changes. This is certainly true for some investors, such as day traders, market makers, and specialists, but the vast majority of investors react only with considerable delay. If one examines the amount of new money invested in equity mutual funds one is surprised to observe that 2000 was surpassed only by 1996

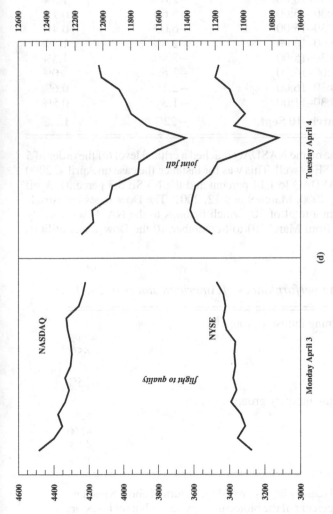

Fig. 5.3d. Short-term behavior of investors on the NYSE and NASDAQ: the switch–sell pattern

Notes: The left hand scale corresponds to the NASDAQ Composite index, while the right-hand scale corresponds to the Dow Jones industrials index. As long as the fall of the NASDAQ remained below a critical level f_c of the order of 5 percent, investors sought shelter in the 30 blue chip companies of the Dow Jones. However, when the fall on the NASDAQ exceeded f_c the decrease spread to blue chip companies as well: between 9:30 and 13:00 the NASDAQ composite lost 13.9 percent while the Dow Jones lost 5.3 percent (a ratio of 5.3/13.9 = 0.38). The same succession of events, that is a flight to quality followed by a joint fall (referred to as the switch–sell pattern), occurred repeatedly in 2000 and 2001; for instance on April 13–14, 2000 (ratio for joint fall was 5.6/9.7 = 0.58), May 2–3, 2000 (ratio was 1.7/2.4 = 0.71, note that the fall of the NASDAQ was limited only by a last minute surge), May 8–9–10, 2000 (ratio was 1.9/5.5 = 0.34), March 8–9–12, 2001 (ratio was 1.97/5.35 = 0.37 for March 9 and 4.10/6.30 = 0.65 March 12).

Source: http://finance.yahoo.com

Table 5.4. *Flight to quality: from the NASDAQ to the NYSE*
after the crash of March 2000

Day	Trading interval	NASDAQ Composite	Dow Jones industrials
Mon. 3 April	9:30–16:00	−6.7%	2.8%
Tues. 4 April	9:30–10:00	−2.0%	1.5%
Mon. 10 April	9:30–16:00	−5.6%	0.7%
Wed. 12 April	9:30–15:00	−6.0%	0.4%
Thurs. 20 April	9:30–16:00	−2.3%	1.4%
Mon. 24 April	9:30–15:00	−2.9%	1.5%
Tues. 2 May	11:00–16:00	−3.6%	0%
Mon. 8 May	9:30–16:00	−2.1%	0.6%
Tues. 27 June	9:30–11:00	−1.3%	0.5%
Total	**10 March–10 Sept.**	**−22%**	**12%**

Notes: Whenever the price decline on the NASDAQ reached a critical level (of the order of 5 percent) the fall spread to the NYSE as well. This was for instance the case on April 4, 2000 (between 9:30 and 13:00 the NASDAQ lost 14 percent and the NYSE 5.3 percent), April 14, 2000, May 3, 2000, May 10, 2000, March 9 and 12, 2001. The Dow Jones industrials index contains only one stock (in a total of 30) which belongs to the NASDAQ, namely Intel. In the same time interval from March 10 to September 10 the Dow Jones utilities index rose by 47 percent.
Source: http:/finance.yahoo.com

Table 5.5. *Contrasting performances of American stocks in 2000*

Four of the worst-performing industry groups	
Internet	−74%
e-finance	−65%
Hardware	−43%
Telecom	−35%
Four of the best-performing industry groups	
Biotechnology	62%
Oil	45%
Utilities	43%
Bank	17%

Notes: At the same time Treasury bonds provided a return of about 8 percent; this is far less than the 62 percent of the biotechnology sector but for investors Treasury bills had the advantage of providing a riskless shelter.

and 1997. Thus, in spite of the fact that in 2000 the Dow Jones index lost 6 percent and the NASDAQ Composite 39 percent, not only was there no net outflow but there was a larger inflow than in 1999. In this case there was no flight to safety at all. How can one explain such a surprising behavior? This is the subject of the next section.

3 To sell or not to sell?

[Between March and May 2000] the hugely popular OTC&Emerging Growth Fund [operated by a big fund-management firm] has lost 55 percent of its value; and yet new money continues to be invested in it, albeit at the rate of 50 million dollars a week, instead of some 200 million dollars.

We placed this excerpt from the *Economist* (May 27, 2000) at the beginning of this paper in order to show that even in the age of Internet communication and online trading not all shareholders react to adverse news within days.

It is a common belief that the behavior of shareholders depends upon the direction of price fluctuations: if prices increase they buy, if prices decrease they sell. This belief, however, is based more on "common sense" than on facts. In this section we present evidence for a specific class of shareholders, which shows that the actual behavior of shareholders can be markedly different. For instance, they may continue to buy despite a prolonged fall in prices or they may sell even though prices climb. A closer analysis shows that a substantial proportion of investors are more influenced by the "general social climate" than by actual price changes. The percentage of speculative investors who optimize their portfolio on a monthly basis turns out to be about 5–10 percent.

In order to fulfill our objectives we must focus on a specific class of shareholders. Indeed, to our knowledge, the only published figures regarding flows of money into or out from stock markets are those for investors who own stocks through mutual funds. Subsequently such shareholders will be referred to as mutual fund shareholders or as mf shareholders. The fact that one can know the net flow of money in the weeks, months, or years following a stock price collapse is of central importance for the present study. Subsequently in-flows will be counted positively and out-flows negatively.

To begin with, let us briefly give some general information about mutual funds in the United States. In 1999 there were about 8,000 funds of which bond funds represented 30 percent and equity funds (also called stock funds) about 50 percent (*Mutual Fund Factbook* 2000). The total assets of equity funds represented 2.4 trillion dollars in 1997, that is 25 percent of the capitalization of the New York Stock Exchange (NYSE) or 20 percent of the combined capitalization of the NYSE and NASDAQ (*Statistical Abstract of the United States* 1999). These orders of magnitude show that in the late 1990s the impact of mf shareholders was far from being marginal. Similar figures for previous years are given in table 5.6(a). The mf shareholders are either households or institutions; in 1999 the latter (mainly retirement funds) represented one third of the assets. Table 5.6(b) summarizes net purchases between 1971 and 1999.

Table 5.6(a). *Assets of equity funds (1975–1999)*

	1975	1980	1985	1990	1995	1999
Equity funds (assets, bn dollars)	37	44	117	239	1,249	4,041
NYSE capitalization (bn dollars)	685	1,243	1,950	2,820	6,013	11,491
Equity funds (% of NYSE capitalization)	5.4	3.5	6.0	8.5	20.8	35.2

Notes: There is a marked contrast between the situation in 1980 and 1999: in the first case mutual funds could only marginally influence the level of stock prices; however in 1999 they represented more than one third of the capitalization and therefore they were to a large extent able to influence or even control the level of stock prices.
Sources: Statistical Abstract of the United States (various years); *Mutual Fund Factbook* (2000). Investment Company Institute.

Table 5.6(b). *Net purchases of common stocks by long-term mutual funds*

	1971	1972	1973	1974	1975	1976	1977	1978	1979	1980
Net purchases (bn dollars)	0.38	−1.6	−1.9	−0.28	−0.95	−2.5	−3.5	1.6	−2.8	−1.9

	1981	1982	1983	1984	1985	1986	1987	1988	1989	1990
Net purchases (bn dollars)	−0.42	2.5	14	5.7	8.1	16.4	22	16	1.1	19

	1991	1992	1993	1994	1995	1996	1997	1998	1999	
Net purchases (bn dollars)	41	66	126	116	103	224	188	165	174	

Source: Mutual Fund Factbook (2000).

3.1 Formulation of the problem

Table 5.7(a) gives monthly data for in-flows of new money into equity funds for the years 1997–2000. How should these data be interpreted? This is what we want to discuss in the present section. In order to simplify the discussion we define the following classes of investors.

• The class of what we call speculative investors who optimize their investments on a daily or weekly basis. For instance, if the return of stock markets in a given month is negative while at the same time it is possible to get an annual 6 percent interest on bond markets they will make an arbitrage in favor of the latter and shift a substantial part of their assets to the bond market.

- The class of what we call long-term investors who make their decisions on an annual or multi-annual basis.

In real life there is probably a whole spectrum of attitudes between these borderline cases. It is mainly for the sake of simplicity that we restrict ourselves to these two classes. A convenient way to describe an actual population in such a simplified framework is to consider that it is a mix of speculative and long-term investors. One of the key variables in the present study is the percentage represented by each class and one of our main goals is to estimate it reliably.

Now let us see how the above framework can be put to use for the specific example of the monthly data for 2000. Two observations are in order: (i) For almost all months (the only exception is August) the yields of bonds (either short term or long term) were higher than stock market returns. Thus, if all mf shareholders had been speculative investors they would have channeled their money into the bond market (or toward other lucrative markets, such as the real estate market, which experienced a boom in 2000). In other words, the in-flow of new money would have stopped or would even have become negative. In contrast the evidence shows that there was a total net in-flow of 290 billion dollars for 2000, which proves that some mf shareholders are rather long-term investors. (ii) If one computes the regression between stock price changes ($\Delta p/p$) and net in-flows of money (f) one obtains (correlation is 0.57):

$$f = a\Delta p/p + b \quad a = 1.36 \pm 1.35 \quad b = 25 \pm 7 \tag{3.1}$$

which shows that there is a definite relationship between flows and price changes. Incidentally, it can be noted that whereas the error margin for a is fairly high in the present case (this is because the regression is made on only 10 points) it is somewhat smaller for previous years (see table 5.7(b)). Equation (3.1) shows that the population of shareholders does not comprise 100 percent long-term investors (otherwise f would not change with $\Delta p/p$).

More specifically it is possible to derive an estimate of the proportion k of speculative investors from equation (3.1). Indeed, under the assumption of two classes of investors, the flow f would be given by:

$$f/C = k(\Delta p/p) + (1 - k).1 \tag{3.2}$$

C denotes a proportionality constant between f and the price changes; if $k = 1$ equation (3.2) reduces to $f/C = k\Delta p/p$, which means that the in-flows of money would be completely controlled by price changes; if $k = 0$ they would be independent of price changes.

Table 5.7(a). *Net new money invested in equity mutual funds versus stock price changes, monthly data*

	Jan.	Feb.	March	April	May	June	July	Aug.	Sept.	Oct.	Nov.	Dec.
1997												
NY + NAS (%)		−0.23	−4.6	4.3	7.0	4.4	7.6	−3.9	5.8	−3.7	3.1	1.6
Flows ($ bn)	28	17	10	16	20	16	26	14	25	20	18	15
1998												
NY + NAS (%)		7.2	4.8	1.1	−2.6	3.2	−2.0	−16	6.7	7.1	6.2	6.1
Flows ($ bn)	14	24	22	26	18	19	19	−11	6.2	2.4	13	3.2
1999												
NY + NAS (%)		−3.8	4.0	4.7	−2.1	5.3	−3.0	−0.65	−2.3	6.2	4.3	9.3
Flows ($ bn)	17	0.76	12	26	15	19	12	9.4	11	21	18	25
2000												
NYSE (%)		−4.6	9.3	−0.46	−0.15	−0.31	−0.15	5.3	−1.6	0.45	−5.5	
NASDAQ (%)		19.2	−2.6	−15.5	−11.9	16.6	−5.0	11.7	−12.7	−8.2	−22.9	
NY + NAS (%)		4.5	5.0	−5.3	−3.6	5.2	−1.7	7.4	−5.0	−2.0	−9.8	
Flows ($ bn)	41	53	39	34	17	22	17	23	17	19	5.7	
Bond 3-mo. (%)	5.5	5.7	5.9	5.8	6.0	5.9	6.1	6.3	6.2	6.3	5.8	
Bond 10-yr. (%)	6.7	6.5	6.3	6.0	6.4	6.1	6.0	5.8	5.8	5.8	5.7	

Notes: The lines "NY + NAS" give the average price change of the NYSE Composite index and NASDAQ Composite index (weighted for respective capitalization). "Flows" designates the new money invested in equity funds. "Bond 3-mo." and "Bond 10-yr." represent the yield of 3-month Treasury bonds and 10-year Treasury bonds respectively.
Sources: Stock indexes: http://finance.yahoo.com; bond yields: http://www.fms.treas.gov/bulletin; flows: *Mutual Fund Factbook* (2000).

Table 5.7(b). *Regression of monthly flows (f) with respect to price changes ($\Delta p/p$): $f = a\Delta p/p + b$*

Year	a	b	Correlation	Proportion k of speculative investors
1997	0.70 ± 0.50	16 ± 2	0.67	4.1%
1998	0.70 ± 0.97	11 ± 6	0.42	5.7%
1999	1.36 ± 0.61	12 ± 3	0.82	9.9%
2000	1.36 ± 1.35	25 ± 7	0.57	5.1%

Table 5.7(c). *Net out-flows of money (f) caused by stock market crashes: short term perspective, in percent of fund's assets (A)*

Year	Dates of crash	$\Delta p/p$ percent	f/A percent
1987	16 Oct.–20 Oct.	−23	−2.1
1990	15 Jul.–15 Oct.	−21	−1.8
1994	22 March–6 April	−7.0	−0.28
1994	Latin American stocks 15 Dec.–31 Dec.	−15	−1.8
1998	20 Aug.–31 Aug.	−12	−0.46

Notes: The crash of December 15–31, concerned only funds comprising Latin American stocks; it was due to the devaluation of the Mexican peso on December 20, 1994. The linear regression for the five points in the table is $f/A = a\Delta p/p + b$, $a = 0.11 \pm 0.06$, $b = 0.54 \pm 0.36$ (the correlation is 0.90), which gives 17 percent for the proportion of speculative investors. *Sources: Mutual Fund Factbook* (2000), Marcis, West, and Leonard-Chambers (1995), Reid (2000).

Identifying equations (1) and (2) gives:

$$\frac{a}{k} = \frac{b}{1-k} \Longrightarrow k = \frac{a/b}{1 + a/b}$$

With the data for 2000 this leads to $k = 0.054/(1 + 0.054) = 5.1$ percent.

Incidentally, it can be observed that if a/b is small with respect to 1, approximately $k = a/b$.

For other years (table 5.7(b)) estimates for k are in the 5–10 percent range. Such an order of magnitude is consistent with what is known about the behavior of mf shareholders through surveys (see below).

3.2 *Some methodological points*

In the previous section a number of methodological options have been selected
which deserve a more detailed discussion.

1 In considering portfolio optimization we neglected transactions costs. Was this justified?
   For stocks, transaction costs have notably declined in the 1990s; in the late 1990s they were
   of the order of 1 cent per share, that is about 0.2 percent if one assumes the average price
   of a share to be about 50 dollars. For equity funds the decline was much slower however.
   In 1990 transaction costs represented about 1.8 percent and by 1998 they had declined to
   1.3 percent (Reid 2000, 18). Since the return differential between stocks and bonds was
   in 2000 at least 6 percent, even such high transaction costs would not deter speculative
   investors from moving their money from one market to another; however it certainly
   induces much friction and partly explains the reluctance of mf shareholders to switch to a
   different investment strategy. In this respect one should also mention the fact that for pension funds, earnings are tax deferred until withdrawal and that contributions may be tax
   deductible.

2 In so far as equity funds comprise American as well as foreign stocks one may wonder if
   one should not use a world index for stock prices. In fact, throughout the 1990s foreign
   stocks represented only about 10 percent of the total assets of American equity funds:
   the figure was 11 percent in 1990 and 13 percent in 1999 (Reid 2000, 14).

3 One may wonder why the stock prices and money flows in table 5.7(a) were not corrected
   for inflation. However the corrections would be negligible as during the 1990s the annual
   inflation rate was on average less than 2 percent, which means that on a monthly basis
   it was less than 0.2 percent. Such a correction is far smaller than the error margin on the
   flow data. However, when subsequently considering yearly data over the whole decade
   we use deflated figures (over the decade the price increase was about 30 percent).

4 A last point concerns the fact that we used f as our dependent variable instead of
   $\Delta f / f$ or some other measure of fluctuations, such as the difference between monthly
   values and the overall trend (for instance in the form of a moving average). In fact
   the regressions were also carried out for $\Delta f / f$, with basically similar results. We
   preferred using f because this is already an increment: it represents the change in the
   funds assets and $\Delta f / f$ would be a second-order increment. Furthermore, to use the
   difference between monthly values and a trend would mean that we give up trying to
   explain the trend; this would simply disregard the low-frequency response of long-term
   investors.

3.3 *Short-term response (weekly fluctuations)*

Crash of August 1998. Between July 1 and September 30 1998 the Dow Jones
index lost 13 percent. How did mf shareholders react? As can be seen from table
5.7(a) the total in-flow for July–September was markedly lower than in the same
quarter of 1997, but it was still positive. There was only an out-flow of money

in one month, namely 11 billion dollars in August, a reaction which is probably connected to the fact that the most spectacular crash had occurred by the end of August: between August 20 and 31 the Dow Jones index lost 12 percent. However this flow (f) of 11 billion dollars represented only $11/2368 = 0.46$ percent of the total assets of equity funds (A).

Crash of October 1987. During the month of October 1987 the Dow Jones index lost 23 percent; in that month the reaction of mf shareholders resulted in an out-flow of cash representing 3.1 percent of equity fund assets. In November the out-flow was only 0.5 percent and for December the figure was almost the same. It is interesting to observe that 70 percent of the October out-flow occurred in the three crash days of October 16, 19, and 20 (Rea and Marcis 1996). Moreover a survey conducted in November 1987 by the Investment Company Institute shows that only 5 percent of the shareholders had sold shares during the crash. A survey conducted after the more limited crash of March 1994 similarly showed that only a small number (of the order of 5 percent) of mf shareholders had sold shares.

In short, the picture which emerges from the above observations shows that a majority of shareholders are long-term investors who do not care about short-term price fluctuations even when they assume crash proportions. This is in agreement with the opinion expressed by mutual fund managers: "A vast majority of shareholders are seasoned investors who do not intend to redeem [i.e. sell] shares in response to adverse market developments" (Rea and Marcis 1996).

Nevertheless, it would obviously be unreasonable to imagine that mf shareholders will hold their shares no matter what happens in the market. To get a clearer view of this point we now analyze their response to fluctuations extending over one or several years.

3.4 Long-term response (yearly fluctuations)

So far we have focused on weekly or monthly fluctuations, but in the present section we consider fluctuations which have a typical duration of one or two years. Yearly data for the decade 1990–2000 are summarized in table 5.8(a); they are not very conclusive however, for there is almost no correlation between price changes and flows whether expressed in absolute or relative terms. A better procedure consists in slicing the interval 1942–1990 into sub-intervals (of an average duration of about two years) during which there was a steady price increase (or decrease) on an annual basis. For each sub-interval we want to know whether shareholders have sold or bought. The results are summarized in fig. 5.4 and table 5.8(b). "Ideally" we would

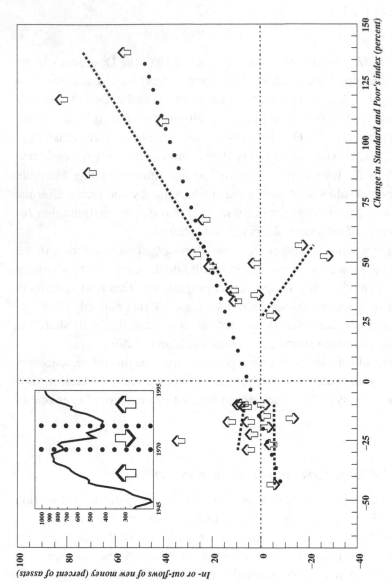

Fig. 5.4. Net new money invested in mutual equity funds as a function of stock price changes

Notes: The dotted line shows the regression line for all points together; it can be seen that there are two clusters of outliers. The arrows indicate the bullish or bearish state of the market, as defined in the inset which shows the evolution of the Standard and Poor's index (a moving average was performed in order to highlight the trends). That additional variable accounts for the anomalous points in quadrants $2(x < 0, y > 0)$ and $4(x > 0, y < 0)$.

Source: Rea and Marcis (1996).

Table 5.8(a). *Net new money invested in equity mutual funds versus stock price changes, annual data*

	1990	1991	1992	1993	1994	1995	1996	1997	1998	1999	2000
NYSE Composite	−7.7	22.5	4.7	7.9	−3.1	31.6	15.3	30.4	16.8	8.3	0.9
Flows, equity, ($ bn)	13	38	74	115	102	107	182	186	127	148	221
Treas., 30-yr, yield (%)	8.5	7.8	7.4	6.5	7.1	6.8	6.5	6.4	5.5	5.8	6.0
Flows, bond, ($ bn)	7	57	66	65	−55	−5	2.5	23	60	4.7	

Notes: The flows are deflated and expressed in 1990 dollars. The regression between price change and flows reads (correlation is 0.22): $f = a(\Delta p/p) + b$ $a = 1.1 \pm 3.1$ $b = 106 \pm 38$ which leads to a proportion of speculative investors $k = 1$ percent.
Sources: Price index: http://minneapolis.org/economy; NYSE Composite index: http://finance.yahoo.com; flows: Reid (2000); bond yield: http://www.fms.treas.gov/bulletin

Table 5.8(b). *Relationship between changes in stock prices ($\Delta p/p$) and the net in- or out-flow of money (f/A) in the time interval 1942–1990: $f/A = a\Delta p/p + b$*

Observations	Number of observat.	a	b	Coef. of correl.	
All data	28	0.32 ± 0.15	5.4 ± 7.3	0.63	(0.34, 0.81)
Bullish climate					
Price increases	10	0.60 ± 0.29	−10 ± 10	0.82	(0.40, 0.96)
Price decreases	10	−0.14 ± 1	5 ± 7	−0.14	(0.69, 0.56)
Bearish climate					
Price increases	4	−0.72 ± 0.79	10 ± 9	−0.78	(−0.99, 0.72)
Price decreases	4	0 ± 0.5	−5 ± 6	0.02	(−0.96, 0.96)

Notes: The observations correspond to successive time intervals during which annual stock prices either increased or decreased. The figures within parenthesis give the confidence interval for the correlation at probability level 0.95. The regression for all data leads to a proportion of speculative investors $k = 5.6$ percent.

again expect a relationship of the form:

$$f/A = a\Delta p/p + b$$

where a is positive and has (more or less) the same value whether $\Delta p/p$ is positive or negative. Such a relationship would mean that mf shareholders buy when prices increase and sell when they fall. However fig. 5.4 shows that the regression line is not the same for positive and negative price changes. For price increases there is an unambiguous positive correlation and regression, whereas for price falls the regression coefficient is almost equal to 0. In other words, mf shareholders responded to price increases by buying new shares, but to price falls they reacted not by selling but by not buying.

One can go a step further. Consider the points in the $\Delta p/p \geq 0$ region. There is an obvious cluster of outliers below the $f/A = 0$ axis, and these points correspond to cases for which shareholders responded to a price increase by selling, rather puzzling behavior. A clue is the observation that all these events occurred during the bear market of 1970–1980. In other words, in an overall bear climate shareholders sell whether the annual price changes are positive or negative.

Is there a mirror effect during bull markets, that is, do shareholders buy during bull markets even when prices fall? We have already seen that this was the case in 2000. In fig. 5.4 the points corresponding to such a behavior would be in the quadrant: $(\Delta p/p < 0, \ f/A > 0)$; not surprisingly all these cases occurred during bull market periods (marked by up-going arrows).

The regression coefficients corresponding to the anomalous cases of quadrants 2 and 3 are defined with large error margins due to the small number of cases in each sub-category; but, at least qualitatively there is no doubt as to the existence of the effects described above. In a sense one is here in the same situation as in neutrino physics in the late 1960s when the number of collision events for an experiment campaign lasting several weeks was of the order of a few dozen. As in neutrino physics, in order to get better accuracy one needs more events.

3.5 Effect of mutual funds purchases on stock prices

So far we have studied how mf shareholders reacted to variations in stock prices; naturally one would also expect that massive purchases of stocks by mutual funds may have an impact on the level of stock prices (remember that by 1999 the assets of equity funds represented as much as 35 percent of the capitalization of the NYSE). From a statistical point of view this is a tricky question however. The standard methodology would be to compute the correlation, with different time lags, between the series of purchases and prices; if prices lagged behind purchases this would be clear evidence that they reacted to purchases. Unfortunately such an effect cannot be observed for the following reasons: (i) Market mechanisms ensure that stock purchases have an almost instantaneous effect on prices (in practice the time lag is less than one minute). Thus, even when using high-frequency data it would be impossible to demonstrate that prices lag behind purchases. (ii) Because purchase movements are fairly smooth, whereas price changes are fairly abrupt, it is the impact of prices which will be predominant in the intercorrelation function $\rho(d)$.

A calculation performed on monthly data between 1997 and 1999 (that is at a time when mutual funds had a strong share of total stock capitalization) confirms this view: the correlation for a zero time lag is 0.47; it becomes negative for any time lag which corresponds to prices lagging behind purchases, but remains positive (albeit small) for purchases lagging behind prices.

Even though it is difficult to identify statistically the effect of purchases on stock price levels, common sense suggests that such an effect exists. Basically one would like to know what price increase results from every billion dollars that mutual funds invested in stocks. A regression performed on the previous monthly data gives:

$$\Delta p/p = af + b \quad a = 0.30 \pm 0.16 \quad b = -2.7 \pm 1.4 \quad r = 0.51 \tag{3.3}$$

p represents the Standard and Poor's 500 index, $\Delta p/p$ is expressed in percent, and the net purchases f are expressed in billion dollars. Equation (3.3) is similar to equation (3.1) except that the roles of the independent and dependent variables are interchanged.

If we assume for a moment that the previous regression indeed describes the actual impact of purchases on prices one is led to the conclusion that every billion dollars invested in stocks leads to an average price increase of 0.30 percent. In order to see if this order of magnitude is plausible let us apply it to the period 1990–1999. During this period mutual funds invested a total of 1200 billion dollars in common stocks, and, according to the above estimate, this should lead to a price increase of $1200 \times 0.30 = 360\%$. As in 1990 the S&P 500 was at the level 330, this would result in a level of $330 \times 4.6 = 1158$ whereas the actual S&P level at the end of 1999 was 1470. Naturally one would not expect mutual funds to account for the total increase; as a matter of fact the 1518 level is probably an overestimation due to the fact that the coefficient a was estimated at a time when the share of mutual funds was higher than during the rest of the decade. Nevertheless, the above calculation suggests that the 0.30 estimate is not completely absurd (it is not ten times too large or 20 times too small).

That order of magnitude can be confirmed by a completely different calculation performed in an innovative paper by Maslov and Mills (2001). Through a high-frequency analysis of the NASDAQ limit order book, these authors were able to find the relationship between excess supply at a given instant and the ensuing price increase in the following minutes. The variation in the price of a given stock is given by $\Delta p = 0.3s$, where p is expressed in dollars and the excess supply s in 10,000 shares. Assuming the average price of a stock to be of the order of 50 dollars that relationship leads to:

$$\Delta p/p = 1.2s$$

where $\Delta p/p$ is expressed in percent and s in million dollars. If one applies this relationship to the whole market (about $N = 5,000$ shares), one gets $\Delta p/p = 1.2s/N$, where $\Delta p/p$ represents the price index variation of the market. Expressing s in billion dollars (instead of millions) and replacing N by 5,000 one gets: $\Delta p/p = 1.2s/5$, that is $\Delta p/p = 0.24s$. The fact that this estimate is so close to our previous 0.30 estimate is of course a pure coincidence; nevertheless the fact that the two figures have the same order of magnitude is reassuring.

3.6 Conclusion

According to economic theory the equilibrium price of a good is determined by the intersection of the supply (S) and demand (D) curves. For stock markets these curves are largely unknown however. On stock exchanges equilibrium prices are determined by computer codes and, for instance on the NYSE in the case of large transactions, by the intervention of the specialist in charge of that specific stock.

Note that the supply and demand curves cannot be reconstructed from the knowledge of the trading volume $V = S + D$ and equilibrium prices. For this purpose one would need to know the bid–ask prices and bid–ask volumes, that is the prices and volumes on the demand and supply sides. These data are not made public however, except indirectly, for instance through an order book.

This is regrettable because models and simulations require specific assumptions to be made about the relationship between excess supply (i.e. $s = S - D$) and the magnitude of the price changes (sign and amplitude). For instance in the Cont–Bouchaud model (Cont and Bouchaud 2000, Stauffer and Sornette 1999) it is assumed that the logarithm of the price changes in proportion to excess supply. In the previous section we proposed a rough estimate of the relationship between price variations and changes in excess supply.

We have seen that there is no simple relationship between price variation and propensity to buy or sell and that the "overall climate" plays a critical role in this respect. The same observation also applies to other responses of shareholders. For instance, the way they react to a fall in interest rates largely depends upon the "overall climate": in October–November 1998 two cuts of 0.25 percent were able to restore confidence and to end the downward spiral; but in early 2001, after a nine-month bear market on the NASDAQ, the situation was very different, and in spite of three 0.5 percent cuts the downward trend persisted.

What should be the next step? Obviously it would not be reasonable to jump too quickly to the conclusion that *all* shareholders behave in the same way as mf shareholders: inductive reasoning based on only one case is fairly hazardous. As we already emphasized, in order to make further progress one needs additional data. It is not impossible that evidence for other classes of shareholders may be available, for instance in the publications of insurance companies or commercial banks. Once one knows the behavior of three or four classes of investors one would be in a much better position to make a reasonable inference about the form of the excess supply function $s = s(p)$.

4 Connection between property and stock markets

Do property crashes have an impact on the movements of stock prices or vice versa? One would expect a fairly close connection between stock markets and commercial real estate, for an increase in profits drives up stock prices and at the same time it boosts the growth of companies and consequently their demand for offices and commercial land. A similar argument applies to residential real estate. If profits and growth rates are high, increasing personal income will boost the property market. However, one expects a looser connection for in this case the link is more indirect and in addition house prices are dependent upon other conditions, such as the demographic situation.

In the light of the above argument one would expect major stock market bubbles to be accompanied by a property peak. It is really surprising, therefore, that the situation of property markets is hardly ever considered by stock market analysts. Perhaps one of the reasons for this is that analysts are mostly concerned with short-term forecasts, whereas changes in property markets affect long-term expectations.

In the present section we examine this question from a statistical point of view, by comparing long-term price series, First of all, however, let us take a look at the interaction mechanisms by focusing on some simple observations. The first one concerns the impact of property crashes on the economy.

4.1 Impact of property crashes on economic growth

As is well known, at the beginning of the 1990s in Japan there was both a stock market collapse and a downturn in real estate prices. The conjunction of these two shocks had a lasting influence on the rate of economic growth in Japan. Stated in this way, this succession of events appears very specific to Japan. Although around 1990 there were property crashes in several industrialized countries (Australia, Britain, France, Sweden, United States) none of these collapses seems to have led to the same outcome as in Japan. This is only partly true however, as will be seen now by considering the case of California.

In 1975 Californian house prices were comparable to the US national average; by 1980 they were 1.6 times higher; by 1989 they were 2.1 times higher (*The Economist*, June 23, 1990). In other words, as for Japan, there was a long and

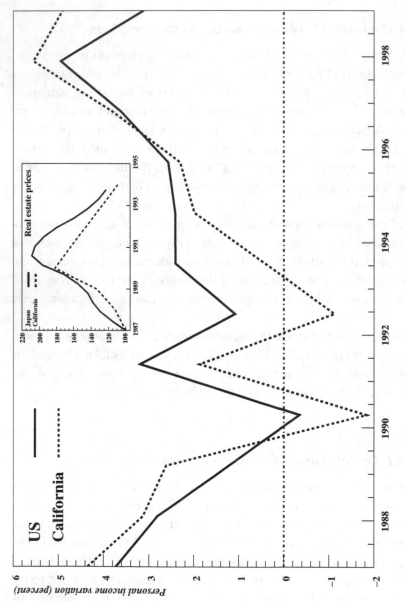

Fig. 5.5. Impact of the real estate crash on the Californian economy

Notes: The insert shows the movement of residential real estate prices in California as compared with Japan. The growth rate of personal income in California dropped sharply in the wake of the property crash and remained below the national average for about six years.

Sources: Real estate prices in Japan: Ramsès (1994); real estate prices in California: *The Economist* (June 23, 1990); personal income: http://www.bea.doc.gov

strong appreciation in property prices. The inset of fig. 5.5 provides a comparison of both markets during the the last stage of the bubble, that is in the two or three years before the downturn. On the same graph I show the variations in national income for California (dashed line) and the United States (solid line). It can be seen that the Californian growth rate dropped under the national growth rate after 1990, i.e. after the property crash, and, more importantly, it remained under the national average until 1996. One can also remember that there was in California a severe financial crisis. It is true that the bankruptcy of the saving banks was partially due to the junk bond crisis, but the depreciation of property assets certainly played a role too – remember in this respect the cascade of bank failures in France in the wake of the property crisis of the early 1990s.

In other words as in Japan the property crash has had a lasting impact on the Californian economy. Naturally for Japan the situation was worse for several reasons: (i) The property bubble was of larger proportions. (ii) The real estate crisis coincided with the stock market crash. (iii) The banking sector of California could get assistance from the Federal government and eventually its economy was pulled forward by the dynamism of the rest of the United States; Japan could benefit from none of these factors. In short, it is reassuring to observe that, in spite of first appearances to the contrary, same causes indeed lead to similar consequences in different countries.

In the next section we emphasize another feature of the link between property and stock crashes which seems to have fairly universal validity.

4.2 Delay in the response of real estate markets

In a general way real estate markets react to business fluctuations with considerable delays; this can be illustrated by the following episodes.

1 The first example is the case of Japan in the 1990s that we have already mentioned: there was both a stock market collapse and a downturn in the real estate market, but the decrease in equity prices was far more abrupt. The downturn in real estate occurred about 15 months after the stock market crash. The same phenomenon was observed in 1882 in Paris (fig. 5.7(a)).

2 Between April 1999 and April 2000 there was a 20 percent decline in property prices in San Francisco's Bay Area; the drop followed the market crash of late August 1998 by seven months.

3 In September 2000 San Francisco gained the dubious distinction of having the most expensive office space in the United States. In Santa Clara county (Silicon Valley) during the last quarter of 2000 the vacancy rate dropped by 11 percent, to 3.3 percent and lease rates for the best class of office space increased by 11 percent, from 68 to 75 dollars per square meter. Throughout 2000 the price of the median house in Berkeley increased by 18 percent (University Wire, February 27, 2001, Cornish and Carey Commercial, *Silicon Valley Quarterly Newsletter*, Fall 2000). These figures suggest that six months after the crash of Internet stocks commercial property in Silicon Valley was still booming.

In short, one observes that the real estate market, whether residential or commercial, reacts with a delay of several months to new economic conditions, whereas stock markets react almost instantaneously.

4.3 The connection between property and stock bubbles

While share prices are highly volatile and unpredictable, property prices are much smoother and, one can expect, easier to understand. For instance, in order for the office vacancy rate to decrease from 30 to 13 percent it is estimated that two years will be required under normal growth conditions and of course even longer in recession times. There are no similar rules for stock markets. This simplicity expectation is an important motivation for studying the link between stock and property markets.

In the following sections we examine long-term empirical evidence for the United States and France.

4.3.1 United States

In terms of capitalization the American property market is more important than the equity market; in 1990 real estate assets were estimated at 8,700 billion dollars (*Le Monde*, March 10, 1992), while the total market value of the shares listed on the NYSE was 2,820 billion dollars (*Statistical Abstract of the United States* 1995). Fig. 5.6(a) provides a comparison between the level of building activity (thick solid line) and of stock prices (thin solid line) – we are mainly interested in a comparison of the peaks and troughs and, thus, the section of the curves corresponding to the period 1875–1915, during which there was no clearly defined peak, can be left out. By visual inspection it can be seen that the peaks and troughs occur almost at the same time in both markets, or at least within an interval of two years. A noteworthy exception to that rule is the stock price peak of 1929 as in this case the real estate activity had already peaked around 1925. In most cases stock prices reached their peak (and trough) *before* the index of building activity. For the time interval 1830–1875 the correlation is 0.30, but it is, of course, dwarfed by the fact that there is an up-going trend in the stock price curve; if one takes the correlation of first differences in order to eliminate the trend the correlation becomes 0.38.

Fig. 5.6(b) provides similar evidence for the period 1963–1999. The overall correlation is 0.51 but as there are no clearly defined peaks for that time interval one is not in a very good position to analyze the interdependence between speculative episodes. Note also that for the United States there is no global property market but rather three regional markets: the North-East (Boston, New York), California (Los Angeles, San Francisco), and Texas (Dallas, Houston). Usually these regions do not peak simultaneously, a circumstance which avoids global downturns.

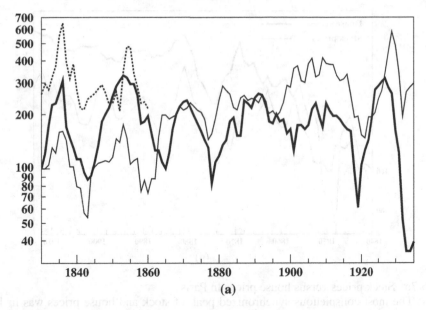

Fig. 5.6a. Stock prices versus building activity in the United States
Notes: Thick solid line: Warren and Pearson index of building activity; dashed line: public land sales; thin solid line: stock prices. Between 1830 and 1880 there were three peaks and troughs and in each case real estate speculation was more or less synchronized with high stock prices (the correlation is 0.30); the connection became looser after 1880.
Sources: Warren and Pearson (1937), Cole (1927), Cole and Frickey (1928).

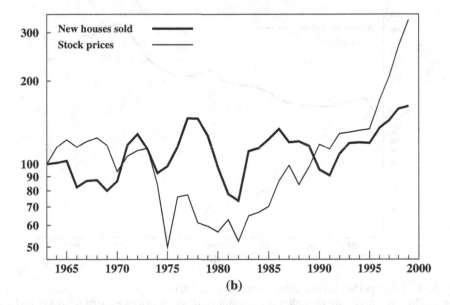

Fig. 5.6b. Stock prices versus number of houses sold in the United States
Notes: The correlation is 0.51.
Sources: Historical Statistics of the United States (1975), http://www.census.gov, http://finance.yahoo.com

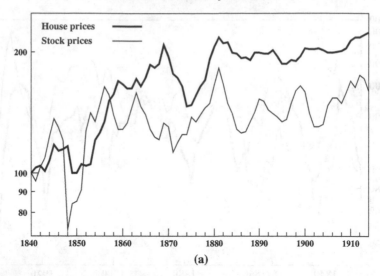

Fig. 5.7a. Stock prices versus house prices in Paris
Notes: The most conspicuous synchronized peak of stock and house prices was in 1882:
note that the fall in house prices was much slower than the fall in stock prices; the same
observation can be made between 1900 and 1902.
Sources: House prices: Documents sur le problème du logement à Paris (1946). Ministère
de l'Economie. Imprimerie Nationale. Paris; Reznikov (1990); *Annuaire Statistique de la
France* (Rétrospectif 1966). I am grateful to Emmanuel Doumas for bringing the source
about house prices in Paris to my attention.

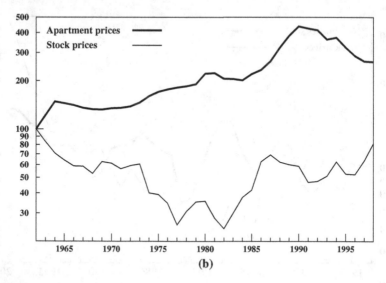

Fig. 5.7b. Stock prices versus apartment prices in Paris
Notes: The two curves are completely disconnected; stock prices experienced a two-decade
long decline from 1961 to 1982.
Sources: Ansidei, Carassus, and Strobel (1978), *Le Marché Immobilier Français: Les
chiffres et les sources* (1993); *Chambre des Notaires de Paris*; Roehner (1999); *Annuaire Statistique de la France (Rétrospectif 1966)*; *Main Economic Indicators. Historical
Statistics*, 1966, 1980; http://finance.yahoo.com

4.3.2 France

Fig. 5.7(a), (b) concern the case of France; fig. 5.7(a) gives the price of houses in Paris. The most spectacular synchronized peaks for house and stock prices was 1881; as one knows the bull market of the 1870s was brought to an end by the sudden crash of January 1882. Fig. 5.7(b) shows a completely different picture in the sense that there is no connection whatsoever between the price of apartments in Paris and the level of stock prices.

6

Speculative peaks: statistical regularities

"What goes up must go down." That was the title of an article published in *Newsweek* on December 16, 1996 which referred to stock market prospects. This warning was of little use to investors however, for one never knows when the downturn really will occur; in this case the bull market went on for about three years. In the present chapter, through a systematic statistical analysis of speculative peaks, we put forward a weaker, but nevertheless useful, form of the previous adage, namely: "The way it goes up, the way it goes down." In other words, once the market downturn has occurred one can with a reasonable level of confidence predict to what level it will fall back. More generally in this chapter we present a number of statistical regularities observed during speculative bubbles. First, we analyze an indicator that can be used to assess the temperature of speculative frenzy; secondly, we show that speculative price peaks follow a definite pattern that we call the sharp peak–flat trough pattern; thirdly, we study the short-term phenomena connected with stock market crashes; and, finally, we review some of the economic consequences of market collapses.

1 A "thermometer" of speculative frenzy

In an article that appeared in 1996 in the *Financial Times* (May 19) Barry Riley wrote: "The [American] stock market's capitalization has risen to almost 90 percent of GDP, beating the previous peak of 82 percent in 1929. The long-run average has been 48 percent." This observation implicitly suggests that the ratio of stock capitalization to Gross Domestic Product can be a reliable indicator of "speculative temperature." But such a proposition can hardly be taken seriously for there has been a long-term upward trend for that ratio. Moreover, the forecasts published by analysts have not been able to identify the market's turning point as is shown in table 6.1.

A major speculative bubble is a phenomenon that is not restricted to financial circles but affects the whole society. Therefore many social indicators could qualify as

Table 6.1. *Brokers' forecasts and recommendations for Cisco Systems*

	March 30 2000	April 30 2000	May 30 2000	Dec. 16 2000	Jan. 16 2001	Feb. 16 2001	March 16 2001
Strong or moderate buy	97%	97%	97%	94%	95%	91%	74%
Strong or moderate sell	0%	0%	0%	0%	0%	0%	0%
Share price ($)	73.6	69.3	59.9	48.2	38.5	28.2	19.9

Notes: The recommendations are an average for a panel of brokers (numbering between 18 and 38 depending on the date). Cisco Systems is one of the largest companies in the Internet industry. It offers hardware, software, and networking services to individuals and firms; by March 16, 2001 its capitalization was 145 billion dollars which made it one of the most important companies of the NASDAQ market. Between November 10, 2000 and March 16, 2001 its price–earnings ratio fell from 130 to 50. In spite of a sharp downturn of the NASDAQ in the months following March 2000, recommendations changed very little: the only shift was a small increase in "hold" recommendations (26 percent in March 2001). Cisco is but one case among many similar ones. For Oracle the gap between forecasts delivered by the company and actual performances was so outrageous that shareholders initiated a securities fraud class action suit against its chairman.
Sources: Business Wire (March 16, 2001), http://finance.yahoo.com.

thermometers of speculative frenzy or, in terms that became popular after the publication of Robert Shiller's book (2000), as an exuberance index. In previous publications (Roehner and Sornette 2000, Roehner 2001a) we considered for instance the annual number of books published on the question of speculation or the number of articles published on that topic in newspapers. These indicators are very "coarse grained" however, and in addition there is a delay between the indicator's movement and the bubble (that delay is of the order of one year for the book indicator). In this section I consider what can be called a generalized price–earnings ratio (PER).

The price–earnings ratio, that is to say the ratio of the price of a share to the annual earnings of the company in the current year, is routinely used in order to assess whether a given stock is over- or undervalued. A first generalization consists in using the average PER for a large sample of stocks in order to estimate the situation of the whole market. Such global PERs data were brilliantly used by Robert Shiller (2000) in his seminal book in order to get a fairly objective assessment of the market's situation.

In fact, the PER concept has a wider domain of applicability; it can be defined for any asset which provides annual returns. For a house or an apartment it will be defined as the ratio of the price to the annual rent; for a bond it will be the ratio of the bond's price to the annual amount of coupon interest (one may remember that the so-called coupon interest is the interest announced when the bond was issued).

For stocks the average historical PER level is somewhere in the 15–22 interval. What is the significance of such a figure? Assuming the price of the asset to be more or less stationary over the course of time, this means that the owner of the asset will cover the purchasing expense in 15–22 years. This is a little less than the time span between two generations or also one half of the average duration of professional life. In other words one can posit that the 15–22 range has a definite meaning, which is related to human life expectation. If, like some tortoises, men were to live 400 years one may think that the historical PER level would be substantially higher.

If this conjecture is correct the historical PER for houses, apartments, agricultural land, or bonds should be in the same range as for stocks. Is this the case? In the following paragraphs we examine the cases of the real estate and bond markets.

1.1 Real estate

As a first indication one may mention the case of Paris between 1985 and 1995. In 1990 at the height of the property bubble the price of a three-room apartment was about 2 million francs (about 300,000 euros or dollars) while the rent was around 6,000 francs per month; this gives a PER of 2,000,000/72,000 = 28. After the bust of the early 1990s the price fell to about 1.5 million while the rent remained basically unchanged which corresponds to a PER of 21. Such movements in PER levels parallel what can be observed in the case of a stock market collapse except that in this case the fluctuations are more pronounced: after a collapse the average PER of stocks can fall as low as 10 or even lower in some cases. The fact that stocks react quicker and with larger amplitudes is a feature that we have observed repeatedly and which will be further investigated in the last part of this book. The previous example is of course not sufficient to draw definite conclusions, but it is difficult to make a systematic study because rent statistics are not commonly published in statistical yearbooks. There is at least one other case for which rent data were available; it concerns Hong Kong in 1998 (first quarter) and the source is the Hong Kong Rating and Valuation Department. The figures for the PER are as follows:

		PER
Hong Kong, center	less than 40 square meters	18
	between 70 and 100 square meters	20
	over 160 square meters	20
Hong Kong, New Territories	less than 40 square meters	21
	between 70 and 100 square meters	24
	over 160 square meters	25

In spite of prices that range from 35,600 HK dollars to 75,500 HK dollars per square meter the PERs are in the same fairly narrow interval (from 18 to 25) already observed for Paris.

1.2 Bonds

The PER of a bond is the ratio of the current price (which is usually close to the nominal price of 100) to the coupon rate (i.e. the nominal interest rate) expressed in percent; in other words it is the inverse of the current yield. Let us see what is the typical order of magnitude of the PER of bonds.

In January 1857, ten months before the banking panic of October, the average yield of a sample of 13 railroad bonds was 8.05 percent (Macaulay 1938), which gives a PER of 12.4. Between January and the height of the crisis in October, the prices of bonds fell from 84.8 to 68.2 because of a marked increase in interest rates (which made previously issued bonds comparatively less profitable); as a result of this price decrease, the PER dropped to 10. In March the crisis was almost over, interest rates plummeted, and the PER of bonds rose to its previous level of 12.5.

Another example is the year 1893, which was marked by a 25 percent drop in stock prices between January and August. In January the average yield of a sample of 37 bonds was 4.44 percent, which corresponds to a PER of 22.5; in August, at the height of the crisis, the PER dropped to 20.6 and in December, when the crisis was almost over, the PER climbed back to 22.4.

In the previous examples the PER of bonds decreased during the crisis. At first sight this could seem paradoxical because bonds provide a safe shelter when the equity market tumbles and one would therefore expect their prices to increase. In 1857 and 1893 this effect was masked by the increase in interest rates, which had the opposite effect. But the effect can be observed during the beginning of the Great Depression of 1929–1931, for in this case interest rates did not increase as in the previous cases but rather declined. In September 1929 the yield of the Aaa grade (the best grade in the investment grade class) corporate bonds was 4.80 (according to a series given on Moody's Website). As the crisis developed the price of bonds increased and the yield progressively decreased to 4.36 in June 1931; this corresponds to an *increase* in the PER from 20.8 to 22.9.

From the above examples one can see that PER levels of bonds are fairly comparable to equity PER levels; furthermore during stock price collapses, if one can set aside other exogenous factors (such as a change in interest rates for instance), the average PER of bonds moves in the expected direction, albeit with an amplitude that is far smaller than the average equity PER.

In short, average PER levels for stocks, houses, apartments, or bonds are approximately in the same range; moreover PER levels increase as a speculative bubble develops and decrease as it bursts. It has been argued by some analysts that in the New Economy the average PER level of 200, reached on the NASDAQ in late 1999, could be as "natural" as any other level. The above data for real estate and bond markets on the contrary suggest that such a high level is at variance with all available evidence, not only for stock markets but for other assets as well. For instance a PER of 200 would correspond to a three-room apartment whose monthly rent was 1,000 euros (a fairly normal rent) but whose price was 2 million euros, which by all standards would be considered as an astronomical price. Alternatively a PER of 200 would correspond to a bond with a coupon rate of 5 percent and which would be priced at 10 (i.e. ten times less than the nominal price); no Aaa bond has ever been priced at such a low level.

What is the role played by PERs during a speculative bubble? During the ascending phase of the speculative episode investors do not care about rents, dividends, or coupon interests; they are interested in capital gains. As a result the high levels reached by the PER remain largely ignored. However, once prices begin to decline, PER levels shift back to the forefront. For companies or property values with large PERs even the most insignificant news may then provide an incentive to sell. During the fall of 2000 this process was quite apparent; the price of stocks with a PER over 100 fell heavily in late 2000 and early 2001: between November 10, 2000 and March 16, 2001 the PER of Cisco Systems fell from 130 to 50; between December 1999 and March 16, 2001 the PER of Oracle dropped from 100 to 13; between October 13, 2000 and March 16, 2001 the PER of Yahoo plummeted from 125 to 31.

2 Shape of price peaks

"How low will the NASDAQ go?" was the title of one of the news items on the Yahoo website on March 16, 2001, and it was indeed a crucial question at that time. The regularities delineated in the present section should help us to answer that kind of question. More specifically, by investigating the shape of speculative peaks we will see that there is indeed a definite pattern. One may wonder if this is not in contradiction with what has been said in previous chapters about the complexity of stock markets and the difficulty of making predictions. An analogy can help us to better understand that apparent paradox.

Traffic jams are fairly unpredictable because they depend upon a large number of factors, e.g. weather conditions, timing in the traffic, highway maintenance, accidents, some of which are completely random. However, once begun, traffic jams display fairly recurrent patterns as to average duration, behavior of drivers, and so on. That traffic jam analogy was introduced by Charles Tilly (1993) in the context of historical sociology, but it applies fairly well to speculative peaks: although it is almost impossible to predict when a speculative peak will start or when the downturn will occur, once the downturn has occurred one can expect that standard patterns will apply.

In order to introduce the topic of this section we present in fig. 6.1 four different price peaks. In terms of amplitude (i.e. peak price divided by initial price) they range from 2 for house prices to about 7 for stamps. In the following paragraphs we

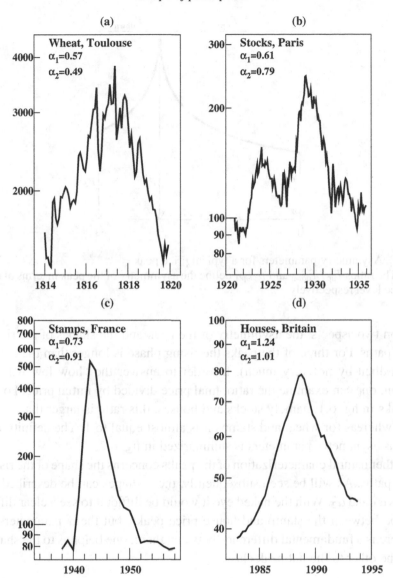

Fig. 6.1. Four price peaks
Notes: All vertical scales correspond to deflated prices. The parameters α_1 and α_2 characterize the shapes of the rising and falling paths respectively; they capture differences in shape
that are not easy to detect with the naked eye; for instance they show that house price peaks
follow a flat peak pattern (the αs are larger than 1), whereas the other peaks follow a sharp
peak pattern (the αs are smaller than 1). (a) Prices expressed in centimes per hectoliter at
the wheat market in Toulouse (south west of France). (b) Stock price index; note that the
downturn occurred in February 1929, that is nine months before the crash at Wall Street.
(c) Prices (expressed in francs of 1910) of a nineteenth-century French stamp (Yvert and
Tellier number 2). (d) Average house prices in Greater London expressed in thousands of
pounds of 1983.
Sources: Drame *et al.* (1991), Hautcoeur (1997), Massacrier (1978), Halifax Group. I am
most grateful to the people at Halifax for their kind assistance.

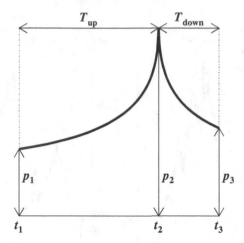

Fig. 6.2. Asymmetry parameters for a typical price peak
Notes: The ratios T_{up}/T_{down} and p_3/p_1 define the asymmetry of the peak in terms of duration and price level respectively.

focus on two aspects: the asymmetry of the peak and the shapes of the rising and falling paths. For three of the peaks the rising phase is longer than the downward phase (albeit by not very much). In order to answer the "how low will it go?" question, one can examine the ratio, final price divided by initial price. For two of the peaks in fig. 6.1, namely stocks and houses, this ratio is larger than 1 (but not much) whereas for wheat and stamps it is almost equal to 1. The definition of the previous asymmetry parameters is summarized in fig. 6.2.

Another natural characterization of the peaks concerns the shape of the rising and falling paths. As will be seen subsequently, these shapes can be described by two indexes α_1 and α_2. With the naked eye it would be difficult to see a clear difference in shape between the stamp and house price peaks, but the α parameters tell us that there is a fundamental difference between them; one belongs to the sharp peak class, the other to the flat peak class.

2.1 *Empirical evidence for asymmetry parameters*

First of all one must explain how the peaks are selected, where does a peak begin, and where does it end? For annual price series the answers can be stated simply. Between t_1 and t_2 (fig. 6.2) there should be only increases, between t_2 and t_3 there should be only decreases; moreover t_1 should be the first increase after a series of decreases, whereas t_3 should be the last decrease before a series of subsequent increases. When using higher-frequency data (quarterly, monthly, or weekly) there will be a number of accessory fluctuations which make the identification of t_1 and t_3 less obvious. To solve the problem one first gets rid of the accessory fluctuations

Table 6.2(a). *Asymmetry parameters for price peaks*

Item	Period	Number of cases	T_{up}/T_{down}	p_3/p_1
Wheat	1486–1913	44	1.4	1.0
Sugar	1962–1981	3	1.5	1.4
Real estate	1900–2000	4	1.3	1.3
French stamps	1935–1949	56	0.78	0.96
Stocks	1870–1993	8	1.3	1.3

Notes: All wheat price peaks in the sample have an amplitude larger than 2. For stamps the 56 cases correspond to different nineteenth-century French stamps. For stocks the cases are detailed in table 6.2(b). None of the asymmetry parameters is less than 0.5 or larger than 1.5 which means that, at least as a first approximation, peak prices can be considered almost symmetrical.
Sources: Roehner (2001a), Massacrier (1978).

Table 6.2(b). *Asymmetry parameters for stock price peaks*

Market	Period	T_{up}/T_{down}	p_3/p_1
NYSE	1903–1907	0.78	1.26
NYSE	1921–1932	3.0	0.83
NYSE	1950–1982	0.96	1.85
Tokyo	1985–1993	1.45	1.25
Berlin	1926–1931	0.29	1.28
Paris	1870–1886	2.0	1.11
Paris	1922–1936	1.21	1.0
Paris	1950–1972	0.96	1.75
Average		**1.33**	**1.29**

Notes: The standard deviation of the T_{up}/T_{down} ratio is 0.84; for p_3/p_1 it is 0.35.
Sources: NYSE, Tokyo: Farrel (1972), Historical Statistics of the United States (1975), *Main Economic Indicators 1969–1988*, OECD; Berlin: *Statistisches Jahrbuch für das deutsche Reich* (various years); Paris: *Bulletin de la Statistique Générale de la France et du Service d'Observation des Prix, Bulletin Mensuel de Statistique*, Hautcoeur (1997).

by using a centered moving average; on the smoothed curve one can then use the same procedure as in the case of annual prices. Naturally, once t_1, t_2, and t_3 have been identified the asymmetry parameters are computed from the primary data, not from the smoothed ones.

The results are summarized in tables 6.2(a), (b). The main conclusion is that price peaks are almost symmetrical: indeed the asymmetry parameters are all within a fairly narrow interval centered on 1. However, one may wonder to what extent that conclusion depends upon our selection procedure; in other words if we selected peaks one should perhaps not be surprised that the rising and falling paths are comparable. This question can be stated in a somewhat equivalent form by asking whether there are many rising paths which are *not* followed by a corresponding fall. For instance, if one considers the course of American stock prices (see chapter 4) one can see that there were only few strong rises which were not followed by a subsequent fall. The major exceptions were the rise of amplitude 2.5 between 1842 and mid 1844 and the rise of amplitude 3.3 between mid 1861 and 1864.

There is a complementary criterion which may help us to distinguish between speculative and non-speculative rising paths. In the first case, when represented on a graph with a logarithmic vertical scale, the rising path is concave on its upper side, which implies an increase which is faster than an exponential and an acceleration of speculation toward the end of the peak; whereas, in the second case, the rising path is linear or even slower. That argument leads us to a closer examination of the shape of the peaks.

2.2 *Mathematical description of the shape of peaks*

For the sake of simplicity let us reason on the logarithm of prices rather than on the prices themselves; furthermore let us assume that $t_2 = 0$ and $p_2 = 1$. On account of what we said previously of the rising path of a speculative peak one will describe it by a function of the form:

$$\ln p(t) = -a|t|^\alpha \quad a > 0, \ \alpha < 1 \tag{2.1}$$

For such a peak the derivative becomes infinite at the peak (i.e. $t = 0$); this justifies the "sharp peak" expression that we have already used.

However, for a peak described by the function:

$$\ln p(t) = -a|t|^\alpha \quad a > 0, \ \alpha > 1 \tag{2.2}$$

the derivative at the peak vanishes; this justifies the "flat peak" expression.

When t_2 is different from 0 and p_2 different from 1, expression (2.1) becomes:

$$\ln p(t) = \ln p_2 - \left|\frac{t - t_2}{\tau}\right|^\alpha \tag{2.3}$$

where the parameter a has been replaced by τ in order to emphasize that it merely describes the time scale.

Note that for $\alpha < 1$ expression (2.3) represents the first term of the function describing a log-periodic behavior namely (Sornette, Johansen, and Bouchaud 1996, Johansen and Sornette 1999b, Stauffer and Sornette 1999):

$$\ln p(t) = A + B|t - t_2|^{\alpha}[1 + C\cos(\omega \ln(|t - t_2|))]$$

While the sharp peak–flat trough pattern describes price peaks at the level of annual fluctuations, the ambition of the log-periodic model is to provide a more detailed description with additional fluctuations at a monthly or even (in the vicinity of the peak) at a weekly level. In some cases this model proved astonishingly successful, as for instance in the predictions published in January 1999 (Johansen and Sornette 1999a) about the subsequent movements of the Nikkei index.

Remark. Troughs can be described in a similar way by using the function:

$$\ln p(t) = \ln p_2 + \left|\frac{t - t_2}{\tau}\right|^{\alpha}$$

more details about troughs, and in particular statistical estimates, can be found in Roehner and Sornette (1998).

2.3 Empirical evidence for shape parameters

Through the substitution of new variables both for p and t, it is possible to replace the non-linear fit corresponding to equation (2.3) by a linear adjustment. In this way α and τ can be estimated through ordinary least squares regression and the quality of the fit will be given by the coefficient of correlation. The corresponding results are given in tables 6.3(a),(b) and an example of an adjustment is shown in fig. 6.3. Real estate peaks are the only ones in the list for which the α parameters are larger than 1; in other words all peaks are sharp peaks except real estate peaks.

Moreover, as a rule of thumb, one can observe that the smaller α_1 the higher is the peak. For stock price peaks the regression between α_1 and the amplitude A reads (correlation is -0.44):

$$A = -a\alpha_1 + b \quad a = -3.6 \pm 5.8, \ b = 7.0 \pm 1.6$$

The error bars are fairly large due to the fact that there are only eight cases.

Apart from stock prices an illustration of the above correlation was the speculative peak of 1979 for gold which resulted in a multiplication of the price by a factor 8; its α_1 parameter (the peak was almost symmetric) was as low as 0.36. Such a low shape parameter means that the increase begins very slowly and accelerates strongly toward the end of the peak. With an α_1 parameter of 0.54 and an amplitude

Table 6.3(a). *Shape parameters for price peaks*

Item	Period	Number of cases	α_1	α_2	α
Wheat	1486–1913				
rising paths		97	0.82		
falling paths		81		0.91	
peaks		44			0.66
Sugar	1962–1981	3			0.87
Gold	1960–1990	1			0.36
Real estate	1900–2000	4	1.20	1.43	1.31
Stamps	1930–1987	6			0.91
Stocks	1870–1993	8	0.86	0.76	0.81

Notes: For wheat prices the selected rising and falling paths had an amplitude larger than 2; the selected peaks had an amplitude larger than 1.5. α is the average of α_1 and α_2.

Sources: Roehner (1989, 2001), Roehner and Sornette (1998).

Table 6.3(b). *Shape parameters for stock price peaks*

Market	Period	α_1	α_2	α	A
NYSE	1903–1907	0.81	0.77	0.79	2.3
NYSE	1921–1932	0.64	0.70	0.67	5.5
NYSE	1950–1982	1.33	0.99	1.16	4.0
NASDAQ	1990–200?	0.54			9
Tokyo	1985–1993	0.89	0.93	0.91	2.5
Berlin	1926–1931	1.22	0.60	0.91	2.3
Paris	1871–1886	0.62	0.89	0.75	1.6
Paris	1922–1936	0.60	0.80	0.70	2.4
Paris	1950–1972	0.81	0.42	0.61	3.8
Average	(without NASDAQ)	**0.86**	**0.76**	**0.81**	

Notes: The coefficients of correlation of the linearized data are between 0.88 and 0.92. Except for Paris (1871–1886) and the NYSE (1950–1982), for which we used annual prices, all the peaks are described by monthly prices. Needless to say using annual data leads to an under-estimation of the amplitude. At the time of writing (Spring 2001) the index α_2 of the NASDAQ case could not be determined with precision: the 12 month estimate $\alpha_2 = 1.04$ merely corresponds to a linear approximation of the peak's vicinity; based on a parallel with similar rising paths (particularly NYSE 1921–1932 and Paris 1871–1886) the most reasonable conjecture is $\alpha_2 = 0.74 \pm 0.04$.

Sources: The data sources are given in table 6.2(b).

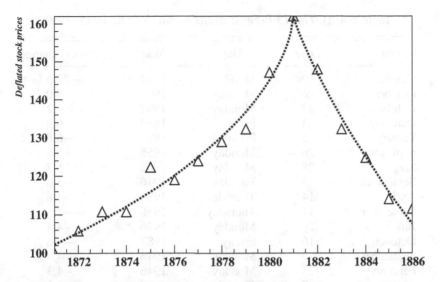

Fig. 6.3. Determination of the shape parameters: example of the crash of 1882 at Paris
Notes: The triangles correspond to annual prices; the dashed line shows the adjusted function estimated through linear regression on the linearized variables. The α index for the rising path is 0.62 (correlation is 0.98) and 0.89 for the falling path (correlation is 0.99).
Source: Hautcoeur (1997).

of about 9, the rising path of the NASDAQ in the 1990s was another example of a highly speculative peak; one may remember that the price increase markedly accelerated at the end of the peak that is to say in late 1999 and early 2000.

In the next section we turn to a closer examination of the turning point of a peak; this leads us to a short-term investigation of market crashes.

3 Stock market crashes

As far as macroeconomic growth is concerned, stock market crashes are unimportant in themselves. The real issue in this respect is whether or not a crash will bring about a shift from a bull to a bear market. If not, as was for instance the case in 1987, the crash will have few lasting consequences. However crashes are of great interest as examples of short-term collective phenomena. From the analysis of stock market crashes one can hope to learn a lot about the behavior of investors in critical moments. In the following sections we examine the standard questions "when?" and "how?" – Note that we do not consider the question "why?"; this would be an ill-defined question as the identification of causes largely depends upon the perspective of the observer. One researcher may look for explicative factors within the exchanges, another will consider the wider sphere of financial markets, and a third will point out macroeconomic factors.

Table 6.4(a). *The 15 biggest drops on the NYSE: 1945–1998*

Month	Date	Day	Year	Percent change
October	19	Monday	1987	−22.6
October	26	Monday	1987	−8.0
October	27	Monday	1997	−7.2
October	13	Friday	1989	−6.9
January	8	Friday	1988	−6.8
September	26	Monday	1955	−6.5
May	28	Monday	1962	−5.7
September	3	Tuesday	1946	−5.6
April	14	Thursday	1986	−4.8
September	11	Thursday	1986	−4.6
June	26	Monday	1946	−4.4
October	16	Friday	1987	−4.6
September	9	Monday	1946	−4.4
February	25	Monday	1946	−4.3
August	27	Thursday	1998	−4.2

Notes: The figures refer to the Dow Jones industrials index; it should be noted that in the late twentieth century it included only one NASDAQ company (namely Intel) and therefore reflected the fluctuations of the NYSE. For the sake of comparison it can be mentioned that on October 24, 1929 the fall was 12.8 percent, 15.8 percent on October 29, and 11.5 percent on October 30. The most crash-prone months are October (33 percent) and September (26 percent). With a percentage of 53 percent (instead of 20 percent in the case of a uniform distribution) Monday is clearly over-represented.
Source: New York Times (August 28, 1998).

3.1 When?

Table 6.4a shows that the months of September and October account for 60 percent of the 15 biggest crashes for the period 1945–1998. If crashes were uniformly distributed these two months would account for only $1/6 = 17$ percent. What factors can explain that concentration of crashes in the fall. Before trying to answer this question it must be emphasized that it is only of anecdotal interest for, if it is easy to list plausible factors, it is almost impossible to really test their impact from a comparative perspective. Among such plausible factors one could mention:
(i) the fact that in the United States mutual funds may sell stocks in October for tax purposes, specifically they try to sell losing stocks to offset gains in other stocks (*USA Today*, October 15, 1999); (ii) companies publish third-quarter earning reports in October, but it is of course not obvious why these reports should be less optimistic than those published in previous quarters; (iii) sometimes the influence of the fall season on the psychological mood of investors is mentioned – the impact of this factor could possibly be assessed by a comparison of stock exchanges located in the northern and southern hemispheres respectively.

Table 6.4(b). *The 10 biggest drops on the NASDAQ: 1972–2001*

Month	Date	Day	Year	Percent change
October	19	Monday	1987	−11
April	14	Friday	2000	−9.7
October	20	Tuesday	1987	−9.0
October	26	Monday	1987	−9.0
August	31	Monday	1998	−8.6
April	3	Monday	2000	−7.6
January	2	Tuesday	2001	−7.2
December	20	Wednesday	2000	−7.1
April	12	Wednesday	2000	−7.1
April	10	Monday	2000	−7.1

Notes: The NASDAQ was created in 1972. April accounts for 40 percent of the cases (instead of 8.3 percent for a uniform distribution); October accounts for 30 percent. As in table 6.4(a) Monday is heavily over-represented: 50 percent instead of 20 percent for a uniform distribution; moreover it can be noted that January 2 was similar to a Monday.
Source: Reuters Securities (January 2, 2001).

3.2 How?

3.2.1 The rebound effect

If the resistance of the air is neglected all falling bodies follow the same law. Does a similar universality exist for falling markets? Well, of course not. However, as fig. 6.4 shows, there is an intriguing similarity between the monthly price series for various crashes. First there is a sharp initial fall which lasts about one or two months and which results in a division of the index level by a factor of the order of 1.40; then there is a second phase marked by a rebound. As it appears from the data in table 6.5 the amplitude of the rebound is determined by the "velocity" of the rise before the downturn.

At this point we are unable to offer any plausible rationale for this rebound effect. One should also note that in some cases the rebound is almost negligible; for instance after the crash of February 1929 in Paris the amplitude of the rebound was only 1.01.

3.2.2 "Frightening Fridays"

During the crash of October 1987 the biggest fall occurred on Monday, October 19: on this day the Dow Jones lost 508 points. However, the week before was already marked by substantial declines. Thus, on Friday, October 16, the Dow Jones abandoned 108 points. As it happens October 16 was the third Friday in October. On March 16, 2001 the Dow Jones lost 2.1 percent while the NASDAQ Composite index lost 2.6 percent; March 16 was the third Friday in March. Is the fact that these

Table 6.5. *The rebound effect after a crash*

Case	Amplitude of rise in the 8 months before downturn	Amplitude of rebound	Time span between downturn and rebound [month]
Paris 1882	1.50	1.21	3.1
New York 1929	1.20	1.15	2.6
New York 1987	1.30	1.11	4.0
Tokyo 1989	1.18	1.07	4.0
NASDAQ 2000	1.74	1.23	3.0

Notes: All stock market crashes considered in this table are marked by a rebound which occurs between 2.5 and 4 months after the crash. The bigger the rise before the downturn, the stronger the rebound (correlation is 0.87). In some cases not listed here there was only a small rebound, as for instance for the crash of February 1929 in Paris.
Sources: See table 6.2(b).

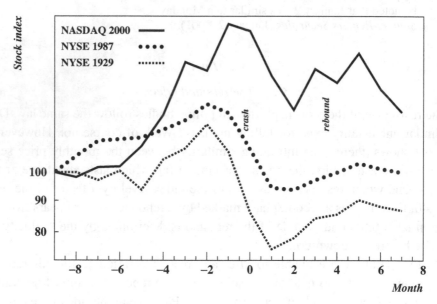

Fig. 6.4. The rebound effect
Notes: Between 2.5 to 4 months after the crash there was usually a rebound. For the sake of clarity we restricted the graph to three cases; two other examples are given in table 6.5.
Sources: Farrel (1972), http.//finance.yahoo.com

strong changes were concentrated on third Fridays merely a coincidence? In this paragraph we explain what made third Fridays so special (a brief explanation was already given in a previous chapter).

In a fascinating book written before the crash of 1987, R. Sobel attracted attention to the fact that third Fridays were marked by exceptional volatility to the point that they were called "Frightening Fridays" by investors. As an illustration let us consider what happened on March 20, 1987, a third Friday. Two hours before the close there was already an unusually high volume of trade as a battle was going on between bulls and bears. Suddenly prices began to sink and when the bell rang the Dow Jones had lost 2 percent in a record volume close to 200 million shares, making it the fourth busiest session in the Exchange's history. What had happened? On third Fridays options on stocks or indexes as well as index futures expire at the end of the session. Thus, in the last hours call option owners (i.e. "bulls") and put option owners (i.e. "bears") would try to push the index up or down to a point where their options can be exercised profitably. A similar phenomenon occurred on June 19, 1987, with a decline of 1.5 percent.

In other words, the fall of Friday, October 16 was the repetition of a type of event that had already occurred recurrently in previous months. In contrast the large decline that occurred on the following Monday was unexpected. In fact, the magnitude of that fall was largely due to a panic created by technical problems: the trading volume was so large that the Exchange was unable to handle it; as a result trade reporting was extremely delayed and traders could not get access to the system (*San Francisco Chronicle*, January 24, 2001).

As already noted in a previous chapter, in normal circumstances options on stocks and indexes have rather a stabilizing effect on underlying stocks. In the years following the 1987 crash "Frightening Fridays" became less frightening. Nevertheless the example of March 16, 2001 shows that the Friday effect was still alive. In the latter case, for obvious reasons, "Frightening Fridays" were also called "triple-witch" Fridays.

3.3 Overnight crashes

In the fall of 2000 huge drops in stock prices occurred recurrently between the close of the exchange and the opening the next day. An example of such an overnight collapse is shown in fig. 6.5 and other instances are listed in table 6.6. Each time the scenario was the same; we illustrate it with the case of Intel. In the late afternoon of Thursday, September 21 (i.e. after the market had closed) Intel warned that its revenue would fall short of Wall Street estimates. Analysts expected a revenue of 1.3 dollar per share whereas the actual revenue would be 4 percent smaller. During the night sell orders accumulated, and, on next day opening, the stock of the world's number one maker of semi-conductors dropped by more than 21 percent; 308 million Intel shares changed hands which set a record for biggest volume in one company in a single day on the NASDAQ. Most of the transactions occurred in the first hour of trading and subsequently the price remained almost flat over the whole day.

Table 6.6. *Examples of overnight falls*

Company	Market	Date	Fall	PER
Intel	NASDAQ	Sept. 22, 2000	21%	
Apple	NASDAQ	Sept. 29, 2000	52%	22
Dell Computer	NASDAQ	Oct. 5, 2000	10%	41
Yahoo	NASDAQ	Oct. 11, 2000	12%	125
DoubleClick	NASDAQ	Oct. 13, 2000	37%	Earnings < 0
IBM	NYSE	Oct. 17, 2000	16%	23
Nortel	NYSE	Oct. 25, 2000	27%	Earnings < 0
Gateway	NYSE	Nov. 30, 2000	33%	12
Apple	NASDAQ	Dec. 6, 2000	16%	7
Microsoft	NASDAQ	Dec. 15, 2000	12%	31
Corning	NYSE	Jan. 25, 2001	13%	152

Notes: The falls occurred after the release of quarterly results. In some cases the results were indeed markedly below expectations, but for some over-valued stocks (e.g. Nortel or Yahoo) the release was merely a triggering factor for a decrease which would bring the price–earnings ratio (PER) in line with the rest of the market; the last column gives PER figures before the price fall.
Source: http://finance.yahoo.com

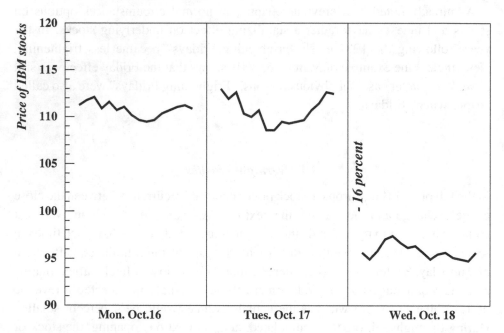

Fig. 6.5. Example of overnight price falls: IBM
Notes: Quarterly earning results were announced on Tuesday evening after the exchanges had closed. During the night sell orders accumulated which led to a sharp drop at opening on the following day. The same pattern was observed for a number of other major companies (see table 6.6).
Source: http:// finance.yahoo.com

This characteristic pattern is illustrated in fig. 6.5 for the case of IBM. Let us for a moment look at this event from the perspective of the NYSE specialist who was in charge of IBM stock. On Wednesday morning before the market opening he had to fix a price level which would balance sell and buy orders. Deciding for a 16 percent price change was certainly not an easy decision; it is clear that many sell orders could not be executed at the price expected by the seller. However the fact that in all these cases the price remained almost unchanged for the rest of the day seems to show that specialists indeed set the opening price at the proper level.

In the next paragraph we consider a question which is of importance for exchange, bank, or government officials in the wake of stock market crashes.

3.4 Lawsuits in the wake of market crashes

Market collapses are usually accompanied by revelations about frauds, by parliamentary inquiries, or by lawsuits. In the search for someone to blame, bribery and fraud practices are revealed which would not have been disclosed or would not have attracted much attention in a more prosperous business situation. A few examples will be enough to give an idea of this kind of recurrent events.

- The 1882 market collapse in Paris was closely related with the failure of the "Union Générale" (see in this respect Bouvier 1960) and a parliamentary inquiry soon exposed extensive fraud in the management of the company.
- In the early 1990s after the failure of the investment company Drexel, Burnham, Lambert, a financial empire based on the junk bond industry, there were many legal complaints. Mr. Milken and other Drexel employees were accused of having obtained funds by using a scheme of coercion, extortion, and bribery. Wilson (1992) quite adequately refers to this episode as "the finger-pointing phase of the business cycle." In 1990 Michael Milken was fined and sentenced to prison for ten years; in 1991 his sentence was reduced to two years plus three years probation.
- In the wake of the market collapse in Japan in the early 1990s there were numerous inquiries: on December 7, 1995 in relation to the failure of the Tokyo Kyowa Credit Association a member of Parliament, Y. Toshio, was arrested; in January 1998 Japan's Finance Minister, H. Mitsuauka, resigned to take responsibility for a bribery scandal; two ministry officials were arrested for having accepted bribes from the banks they supervised.
- In the United States after the 1998 crash and the LTMC (long-term management of capital) failure there were allegations of mismanagement in the press. They were quickly forgotten however, as the bull market resumed.
- In March 2001 the law firm Wolf Haldenstein commenced a class action lawsuit on behalf of all purchasers of Oracle shares during the period between December 15, 2000 and March 1, 2001 (*Business Wire*, March 16, 2001). During this period Oracle chairman, Larry Ellison, sold nearly 900 million dollars of his Oracle stock at a price as high as 32 dollars per share. However, when it became clear that Oracle's 11i e-business suite [a new

business software] was fraught with major technical problems, the shares lost 50 percent to 15 dollars. The complaint maintains that Larry Ellison was fully aware of the problems long before they became known to the public. After Wolf Haldenstein's announcement several other law firms commenced similar class actions against Oracle.

• On March 16, 2001 the law firm of Sirota and Sirota filed a class action lawsuit against several Wall Street investment banks (e.g. Morgan Stanley and Merrill Lynch) who served as middlemen (i.e. underwriters) for the initial public offering of Priceline.com; the lawsuit was on behalf of the persons who purchased the stocks of Priceline.com between March 29, 1999 and March 14, 2001. The complaint states that in order to receive an allocation of shares customers had to agree to purchase additional shares at progressively higher prices, a process know as "laddering a stock". Customers were also required to pay secret commissions to the underwriters which could be as high as one third of the gains made by the customers from the share received in the initial offering. Not surprisingly, such practices sent the stock to artificially high levels. In May 1999 the share price was over 100 dollars, in March 2001 it had dropped to less than 3 dollars (*New York Times*, March 12, 2001). It is typical that such practices were revealed only after the sharp stock price drop of March 2001, that is two years after the events had occurred.

Apart from price fluctuations, movements in trading volume are an important factor in speculative bubbles; big trading volumes improve the liquidity of the market and boost the revenue of brokers, specialists, or market makers. This is the topic considered in the next section.

4 Trading volume

"It takes volumes to make prices move" is one of the adages of investors about trading volume. But how much volume does it take to make prices move by 1 percent (for instance)? This is the question we now examine.

4.1 Volume at the level of individual stocks

If during successive days price variations are in opposite directions the overall price change over a week or a longer time interval may be close to 0 while trading volume in contrast increases with the length of the time interval. As a result, in order to get a reliable estimate of the relationship between volume and price changes one must analyze changes over the shortest possible time interval. In our case the shortest time interval permitted by the data set at our disposal was one day. The results for five different companies are summarized in table 6.7. These companies were selected on two criteria. First, we wanted a broad range of capitalizations: from Pets.com to IBM, capitalization varies from 3 million dollars to 148,000 million dollars, but as can be seen from table 6.7 the magnitude of the capitalization does

Table 6.7. *Relationship between trading volume and price variations*
$$\ln v = a(\Delta p / p) + b$$

Company	Market	Period	$|a_-|$	Correl.	a_+	Correl.
IBM	NYSE	May 15, 2000–Sep. 6, 2000	17	(0.50)	11	(0.60)
		Sept. 7, 2000–Dec. 29, 2000	11	(0.71)	7.6	(0.43)
Yahoo	NASDAQ	May 17, 1999–Dec. 31, 1999	8.8	(0.37)	7.1	(0.55)
		March 2, 2000–Jun. 23, 2000	6.5	(0.48)	4.8	(0.41)
Doubleclick	NASDAQ	Nov. 2, 1998–Nov. 18, 1999	5.7	(0.42)	5.4	(0.54)
		April 3, 2000–Jul. 26, 2000	7.4	(0.55)	5.8	(0.57)
Webvan	NASDAQ	Dec. 1, 1999–Mar. 28, 2000	8.7	(0.51)	6.3	(0.58)
		July 24, 2000–Nov. 13, 2000	5.1	(0.51)	4.0	(0.50)
Pets.com	NASDAQ	May 1, 2000–Aug. 22, 2000	8.5	(0.60)	4.8	(0.45)
		Average (NYSE)	**14**		**9.3**	
		Average (NASDAQ)	**7.2**		**5.4**	

Notes: We used the logarithm of the volume in order to make the regression coefficient independent of possible splits. For instance for Yahoo there was a split on February 14, 2000; for the same reason the selected time intervals do not comprise any split. The regression corresponds to daily data; a_- and a_+ denote the regression coefficients for negative and positive price changes respectively.
Source: http://finance.yahoo.com

not seem to affect the regression coefficient. Secondly, several of the companies (e.g. DoubleClick or Yahoo) experienced very rapid changes in some time intervals and we wanted to see how this affected trading volume. No clear pattern, however, emerged in this respect.

Two conclusions can be drawn from the results in table 6.7. First, one must emphasize the fact that there is a definite relationship between volume and price changes; on account of the erratic behavior of trading volume this came rather as a surprise. Second, the regression coefficient is clearly larger for negative price changes than for positive ones; this is true not only for the averages but also at the level of each separate company. In words, this means that a 1 percent price fall will give rise to a larger volume than a 1 percent price increase. At this point we have no explanation for this phenomenon.

4.2 Volume movements at market level

Another familiar Wall Street adage is that "volume is relatively heavy in bull markets and light in bear markets" (Karpoff 1987, 117). In two previous publications (Roehner and Sornette 2000, Roehner 2001, ch. 5) we have analyzed volume movements for stock as well as real estate markets and we arrived at the conclusion that large volumes and high prices go hand in hand, whereas bear markets are associated with declining volumes. However, after its downturn in March 2000 the NASDAQ market provided what may be seen as a perfect counter-example. Indeed, between March 2000 and January 2001, the NASDAQ Composite index dropped from 5000 to 2500, while at the same time the average daily volume increased from 1.8 billion shares to 2.4 billion shares. In our opinion this does not refute the previous regularity however: volume becomes smaller during a bear market only if all other conditions remain unchanged. As will be seen below, the NASDAQ example violated that "ceteris paribus" requirement. Indeed a revolution in trading was brought about by three factors (i) the introduction of electronic trading, (ii) the growing role of day traders,[1] and (iii) the shift to decimal prices.

4.2.1 Day traders

In early 1997 online equity trades represented 7 percent of all equity trades; by the end of 1998 that proportion had almost doubled (Credit Suisse First Boston). Online trading certainly represents a huge productivity increase especially if orders can be emailed directly to market specialists without having to pass through brokers; this was indeed made possible by the direct-access trading systems (such as Tradescape for instance) used by day traders. In the United States a pool of about 50,000 semi-professional day traders in early 2001 accounted for 81 percent of the retail trading volume (*Fortune*, February 19, 2001, *USA Today*, March 16, 2001). While in the wake of the NASDAQ downturn of March 2000 many online brokers saw a sharp decrease in the trade generated by the occasional traders, which formed the bulk of their customers, the volume generated by day traders markedly increased. This trend was strengthened by the decimalization revolution.

4.2.2 Decimalization

Decimalization, that is to say the shift from prices expressed in 1/16 of a dollar to prices expressed in dollars and cents, was introduced on the NYSE in January 2001 and on the NASDAQ in April 2001. At first sight it could seem to be a purely technical event, but in fact it had far-reaching consequences. One must remember that retail investors buy from middlemen (i.e. specialists on the NYSE or market makers on the NASDAQ) at the so-called ask price, and they sell at the so-called bid price, which is always lower than the ask price – the ask–bid difference (also known as the spread) goes to the middleman. The situation is very much the same as for stamp collectors – the price at which a collector may sell a stamp to a stamp trader is always lower than the price at which the trader will sell it subsequently, and similarly on stock markets . On the NYSE before decimalization the minimum

---

[1] Day traders are short-term traders; depending on their strategy they hold their shares for a period of time that can range from a few minutes to a few days. Since their revenue on each trade is very small the level of transaction costs is of great importance to them.

spread was 1/16 of a dollar, that is 6.25 cents. After the introduction of decimal prices the minimum spread became 1 cent, but it should be noted that the specialist may set it at a higher level.

The spread represents a source of revenue for specialists and market makers but for shareholders it represents an additional transaction cost; and since, after their sharp decline in the 1990s commissions represented no more than 1 cent per share, the spread had become by far the largest component of transaction costs. In other words decimalization opened the door to a sharp decline in transaction costs, which, not surprisingly, would bring about a substantial increase in trading volume. In short, the volume trend on the NYSE and NASDAQ in the early 2000s can be seen as determined by two opposite effects: the shift from a bull market to a bear market which tends to lower trading volumes and the reduction in transaction costs which tends to increase them. It is very difficult to say which of these effects will prove predominant. The main question is how specialists will accept a substantial erosion in their revenue in a bear market environment.

In conclusion, we have seen at least qualitatively that the continuation of the upward trend in trading volume on the NASDAQ even beyond the market downturn does not necessarily contradict the general rule that trading volumes are lighter in a bear market.

5 Economic consequences of stock market collapses

In January 2001 the capitalization of the NYSE was about 11,000 billion dollars; as a result every 10 percent fall would result in the evaporation of 1,100 billion dollars which represents about 4,000 dollars per American.[2] Such a figure helps us to understand why a major stock market collapse has an impact on many, if not all, economic sectors. In this section we will restrict ourselves to three issues: first, we consider how a market collapse affects the level of consumer confidence; secondly, we study the relationship between stock price levels and commission rates; thirdly, we examine to what extent a protracted recession brought about by a market collapse may influence the distribution of income.

Throughout this section one should bear in mind the distinction between market crashes and market collapses. A crash is a sudden price fall in a couple of days or weeks, while a collapse refers to a bear market which lasts at least several months and possibly a couple of years. A crash does not necessarily lead to a collapse nor is a collapse necessarily preceded by a spectacular crash. An illustration of the first assertion was the crash of 1987, and an example of the second was the Tokyo stock market collapse in the early 1990s, which was a long slide rather than a sudden

---

[2] The tax reduction of 1,600 billion dollars over ten years decided by President G.W. Bush would be obliterated by a fall of 14 percent of the NYSE.

crash. As a matter of fact, although chronologically related, these two phenomena are almost completely different. A crash is a microeconomic phenomenon, which results from a panic among investors; on the contrary a collapse is a macroeconomic phenomenon, which results from a downward economic spiral.

One would expect the effects of a collapse to depend upon the capitalization of the stock market. This can be illustrated by the stock market collapse that occurred in France between 1962 and 1981. The SBF (Société des Bourses Françaises) stock price index fell from 108 to 22 but without much prejudicial consequence for the economy. This can certainly be attributed to the market's small capitalization: in 1962 the capitalization of the Paris equity market represented only 30 percent of GDP (Gross Domestic Product); by 1980 its share of GDP had fallen as low as 9 percent (*Annuaire Statistique de la France*).

5.1 Consumer confidence

To what extent is consumer confidence influenced by the performance of stock markets? This is the question that we will examine next. However consumer confidence is of little interest in itself; it is through its possible connection with the business situation that consumer confidence is of significance. This leads us to briefly review to what extent consumer confidence influences consumption.

It is tempting to assume that a high level of consumer confidence goes with a strong spending propensity, while in contrast during a climate of uncertainty and low consumer confidence consumers tend to postpone their spending. The fact that these assertions sound fairly plausible does not imply that they are true however. Let us examine this matter more closely. To begin with, we recall how consumer confidence is estimated.

5.1.1 Consumer confidence estimates

In the United States there are two main confidence indexes and both are estimated through surveys: (i) The consumer confidence index published by the Conference Board is based on about 5,000 questionnaires mailed to a nationwide representative sample of households. Each month a different panel is surveyed; on average about 70 percent of households respond. The survey comprises five questions about the business, employment, and income situations, and for each question there are three response options: positive, negative, or neutral. The survey data are available bi-monthly from 1967 through 1977 and monthly thereafter. (ii) For the consumer sentiment index of the University of Michigan the survey is conducted by telephone on a sample of at least 500 persons; it contains as many as 50 questions. The index data are released only to subscribers.

As a preliminary test it can be interesting to check how closely these indexes are in accordance one with another. From a correlation analysis of quarterly data for two periods for which data are available one gets the following results:

	Period	Correlation	Confidence interval
Consumer sentiment / consumer confidence	1978–1987	0.71	(0.51, 0.84)
Consumer sentiment / consumer confidence	1996–1998	0.98	(0.94, 0.99)

In the United States there is also a producer sentiment index called the "Purchasing Managers Index" (PMI); a correlation analysis over the period 1996–2000 shows that the PMI is ahead of the consumer confidence index by about five months (the correlation is 0.57).

5.1.2 Consumer confidence and consumption

In order to examine the relationship between the level of consumer confidence and the business situation we compare the time series of housing starts with the consumer sentiment (fig. 6.6(a)). Remember in this respect that the housing market is a major component in national expenditure. As can be seen the two curves have a similar shape (the correlation is 0.38). We are particularly interested in the time lag between the two series; from mere inspection of the curves the answer is not obvious for in some time intervals (as in 1980–1982) the confidence sentiment precedes the housing starts and in others (as in 1984–1985) it lags behind housing starts. To get a more quantitative estimate we compute the inter-correlation:

$$\rho(d) = \frac{E[(Y_1(t) - \overline{Y_1})(Y_2(t + d) - \overline{Y_2})]}{\sigma_1 \sigma_2}$$

$\rho(d)$ is maximum for d comprised between one and two quarters:

$$\rho(0) = 0.38, \quad \rho(1) = 0.40, \quad \rho(2) = 0.41, \quad \rho(3) = 0.39, \quad \rho(4) = 0.32$$

which means that on average housing starts precede the movements of consumer sentiment by about four to five months. Intuitively that conclusion is rather unsettling. One would rather expect changes in consumer sentiment to precede changes in housing starts; after all the decision in favor (or against) building a new house is usually based on the expectations prevailing at the moment when the decision is taken. This anomaly is related to the fact that the consumer sentiment index is a broad index based on expectations about various aspects of the national economy. If one focuses on real estate by asking the question, "Do you think now is a good time or a bad time to buy a house?", it can be observed that the answers anticipate

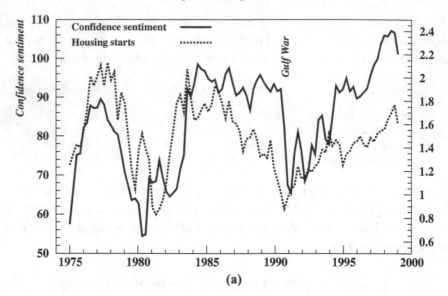

(a)

Fig. 6.6a. Relationship between housing starts and confidence sentiment
Notes: Confidence sentiment lags behind housing starts: the inter-correlation for a lag of
one semester is 0.41.
Sources: Survey of Current Business (April 2000), Michigan University http://athena.
sca.umich.edu

changes in home sales by six months (see in this respect: Survey of consumers,
http://athena.sca.isr.umich.edu).

Now that we know that the consumer sentiment index bears a fairly close relation
to the business situation it is time to examine how it is affected by changes in the
level of stock prices.

5.1.3 How stock prices affect consumer confidence

In fig. 6.6(b) we compare changes in consumer confidence and changes in the Dow
Jones index, on the one hand, and the NASDAQ Composite index, on the other
hand. It can be seen that the consumer confidence index lags behind stock prices.
It is of interest to gauge the respective impact of each market. By computing the
inter-correlation as explained above one gets:

Index	Maximum of $\rho(d)$	for d equal
DJI / consumer confidence	0.76	2 months
NASDAQ Composite / consumer confidence	0.69	1 month

Not surprisingly we find that it is the NYSE which has the predominant influence: the
correlation is larger and it precedes consumer confidence by two months. Note that
we considered a rather short time interval, namely January 1998 to December 2000,

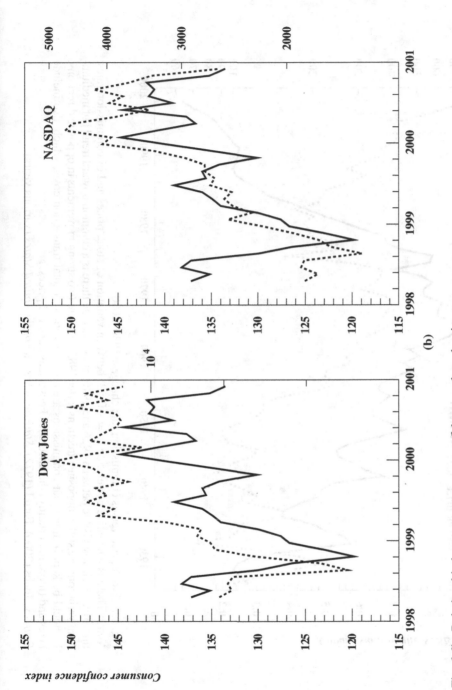

Fig. 6.6b. Relationship between consumer confidence and stock price movements
Notes: Solid line: Consumer confidence index (Conference Board); dashed line: Dow Jones industrials index (left), NASDAQ Composite index (right). Consumer confidence movements follow stock index changes.
Sources: Baron's website, http://finance.yahoo.com

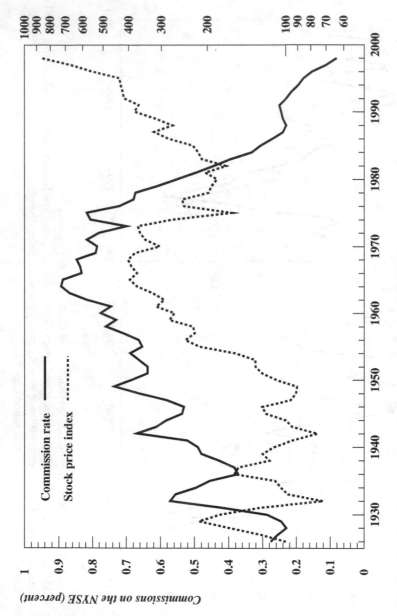

Fig. 6.7. Stock price versus commission rate changes (1926–2000)

Notes: There is a negative correlation ($r = -0.50$) between variation of stock prices and of commission rates; this effect is clearly visible between 1926 and 1940; after that date there is a common upward trend but medium-term variations are still in opposite directions; after 1985 there are two strong movements in opposite directions. It should be kept in mind that commissions are only a fraction of total transaction costs, which also include taxes and (more importantly) the difference between ask and bid prices.

Sources: Jones (2000), Farrel (1972), *Statistical Abstract of the United States* (various years).

which however has the advantage that the movements of the NYSE and NASDAQ were not very correlated (as would have been the case during the bull market of the mid 1990s).

5.2 Relationship between stock price levels and commission rates

Intuitively one may assume that commissions on stock transactions affect the market in two ways: (i) within a fixed pool of stock investors a decrease in commission rates tends to increase trading volume; (ii) lower commission rates tend to attract toward equity markets capital previously invested in other speculative markets. Fig. 6.7 gives an overview of the relationship between commission rates and stock price levels. By mere inspection it can be seen that there are two quite distinct intervals. First, after the deregulation of stock markets in 1975 the two variables are negatively correlated for long- and medium-term movements: the correlation for this interval is −0.71. Secondly, in the interval 1930–1975, during which commissions were set by (changing) market regulations, there are two opposite effects: there is a long-term parallel increase in commission and prices (the correlation is 0.64), but, the medium-term fluctuations still show a negative correlation – this is confirmed by computing the correlation of differences, as one gets a negative correlation of −0.48. For variations expressed in percentages the correlations and regressions are almost the same for the two intervals, as shown below:

	Period	Correlation	Regression coefficient
Prices/commissions	1926–1975	−0.48	−0.91 ± 0.47
Prices/commissions	1976–2000	−0.51	−0.91 ± 0.66

One can summarize these results by the statement:

Rule. Between 1926 and 2000 stock prices and commission rates changed in opposite directions: a 10 percent price decrease (increase) lead to a 9 percent rate increase (decrease).

Commission rates are highly dependent on the organization and productivity of the exchanges, and it is therefore remarkable that this rule applies to such a broad period of time, which saw many organizational transformations. It is not unreasonable, therefore, to expect that it will continue to hold in the next decades.

In order to determine if the increases in commission rates are a cause or a consequence of market collapses one would need monthly or weekly data. However the annual data at one's disposal clearly show that stock market collapses and increases in commission rates go hand in hand. By diminishing the liquidity of the market and by reducing the attraction of stock markets in comparison to other assets these higher commissions reinforce the "bearishness" of the market.

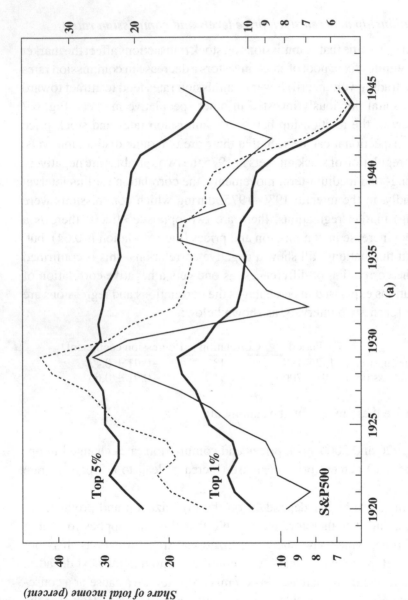

Fig. 6.8a. Impact of a stock market collapse on the distribution of income in the United States

Notes: The right-hand scale corresponds to the Standard and Poor's index. The top 1 percent share of total income is more sensitive to stock price changes (correlation is 0.46) than the top 5 percent share (correlation is 0.30); this suggests that equity revenue represented a sizeable proportion of income only for the most wealthy fraction of the population. The dashed line shows the Gini coefficient.

Sources: North (1961, 1966), Historical Statistics of the United States (1975).

5.3 Effect on the distribution of income

Some decades ago stocks were mainly held by wealthy people, and, therefore, by wiping out a substantial fraction of stock-related income, a stock market collapse could reduce the inequality of income. Is that conjecture confirmed by empirical evidence? This is the question that we now investigate. Unfortunately, at least for the present study, there were not many major stock market collapses and for some of them data for the distribution of income are not available. In the following we consider three cases: the United States before and after 1929, the United Kingdom after 1972, and Japan before and after 1989.

5.3.1 United States (1920–1945)

Fig. 6.8(a) shows the shares of national income of the top 1 percent and top 5 percent. The first curve follows the Standard and Poor's 500 index with a correlation of 0.46; the connection between this index and the second curve is looser (the correlation is 0.30). The fact that the income changes follow stock prices is particularly apparent after the upturns: the income curves reacted with a delay of one or two years.

5.3.2 United Kingdom (1970–1980)

The collapse of the London stock market between 1972 and 1975 was of greater magnitude than the fall in 1929–1930. Fig. 6.8(b) shows that the share of net worth held by the top 1 percent of wealth holders follows fairly closely the movements of stock prices (the correlation is 0.88). Note that in this case the collapse was not due to the burst of a speculative bubble (there was only a modest price increase before 1972), it was rather the combined effect of a number of unfavorable factors.

5.3.3 Japan (1976–1998)

As shown in fig. 6.8(c) the Gini coefficient (g) of the income distribution follows fairly closely the movements of stock prices. One could argue that in the early 1990s there was also a decrease in real estate prices; however that decrease was much slower than the fall in stock prices and there is little doubt that the sharp drop in 1990–1991 reflects the impact of the stock market collapse. A further confirmation is the small increase in g around 1995 that is to say at a time when real estate prices were still declining.

5.3.4 Comparison

In order to be able to compare the three previous cases one must first convert the top 1 percent shares of income (s) into Gini coefficients. As a first approximation we assume that there is a linear relationship between s and g: $g = \alpha s + \beta$. The parameters α, β can be determined from the two extreme situations of perfect

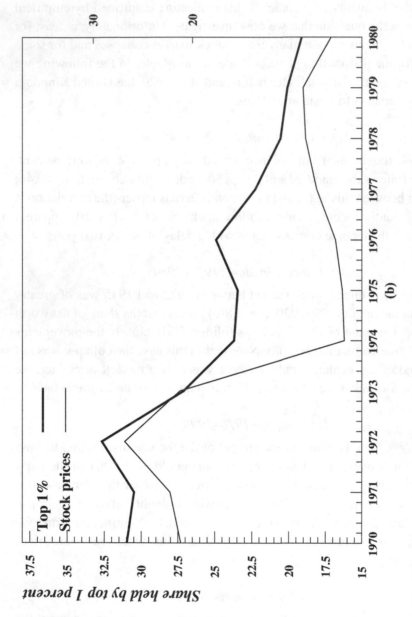

Fig. 6.8b. Impact of a stock market collapse on the distribution of income in the United Kingdom

Notes: The right-hand scale corresponds to the FT500 stock price index. In the UK the crash of 1972 was of greater magnitude than the crash of 1929. The correlation between the top 1 percent share and the stock price index is 0.88.

Sources: Davies and Shorrocks (2000); *Main Economic Indicators: Historical Statistics 1969–1988.* OECD.

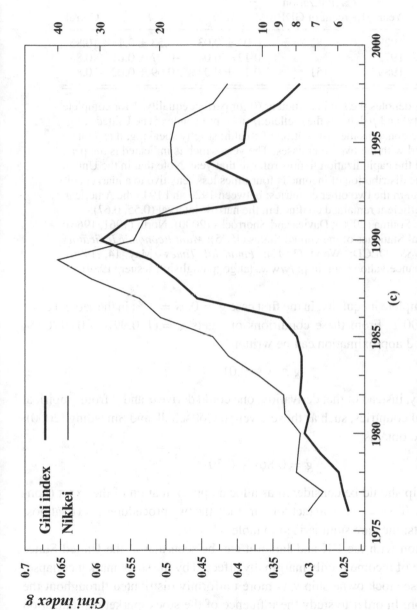

Fig. 6.8c. Impact of a stock market collapse on the distribution of income in Japan

Notes: The right-hand scale corresponds to the Nikkei index. The Gini coefficient is a measure of income inequality which is comprised between 0 (complete equality) and 1 (complete inequality). The correlation between the Gini and Nikkei indexes is 0.85.

Sources: Souma (2000); *Main Economic Indicators: Historical Statistics 1969–1988*. OECD; http://finance.yahoo.com

Table 6.8. *Impact of stock market collapses on the distribution of
income* $g = a \ln p + b$

Country	Year	Capitalization [percent of GDP]	a	b	Correl.
USA	1929	82	0.03 ± 0.03	0.53 ± 0.1	0.85
UK	1972		0.12 ± 0.04	-0.07 ± 0.01	0.88
Japan	1989	151	0.17 ± 0.04	0.09 ± 0.02	0.85

Notes: g denotes the Gini coefficient (0 for perfect equality, 1 for complete inequality) and p denotes the (deflated) stock price index. The United Kingdom case concerns the distribution of wealth; strictly speaking, it cannot be compared with the two other cases. The year which is indicated is the peak year and the capitalization figures refer to that year. Note that in the United States the distribution of income is four times less sensitive to a market collapse than in the two other countries; between 1920 and 1945 the American Gini coefficient remained confined in the narrow interval (0.55, 0.67).
Sources: Souma (2000); Davies and Shorrocks (2000); North (1961, 1966); *Historical Statistics of the United States* (1975); *Main Economic Indicators 1969–1988*, OECD; Wood (1992); *Financial Times* (May 14, 1996); http://finance.yahoo.com; http://www.stat.go.jp/english; Liesner (1989).

equality or complete inequality. In the first case $g = 0$, $s = 1\%$; in the second case: $g = 1$, $s = 100\%$. From these conditions one gets $g = (1/0.99)s - 0.01/0.99$, which to a good approximation can be written:

$$g \simeq s - 0.01$$

Alternatively, instead of that derivation, one could derive α and β from empirical data for several countries, such as those given in Gottschalk and Smeeding (2000). In this way one obtains:

$$g = 0.86s + 0.50$$

that relationship should be considered as a local approximation of the exact (nonlinear) relationship $g = f(s)$. In fact for our study the two procedures lead to almost identical results; they are summarized in table 6.8.

In comparison with the UK and Japan it can be seen that in the United States the distribution of income is only marginally affected by the stock market collapse; perhaps because stock ownership was more uniformly distributed throughout the total population. In order to study the influence of the stock market capitalization one would need additional evidence for other countries. For the sake of comparison it can be noted that in January 2000 the combined capitalization of the NYSE and NASDAQ was 16,500 billion dollars (*New York Times*, March 21, 2000), which represented 170 percent of the GDP (9,752 billion dollars in the first quarter of 2000).

Part IV

Theoretical framework

7

Two classes of speculative peaks

In September 1692 a wholesale trader named Massenot who was a purveyor for the "Marine Royale" (French navy) bought 200,000 pounds of wheat in Burgundy at a price of 11 livre per setier (1 setier = 117 kilogram). This was in a fact a forward transaction in the sense that the wheat was to be delivered one half by November 11 and the remainder by December 25. But in the meanwhile the price of wheat increased markedly and by the end of the year it had reached a level of 20 livre per setier. Subsequently, instead of being used by the navy, the wheat was sold to the bakers of the city of Dijon (north of Burgundy) and as a result Mr. Massenot made a handsome profit of 4,000 livres (Martin 1908). The permission to carry out forward transactions was an important privilege of army and navy purveyors; needless to say, as in the above case, that privilege was often used for making huge profits in times of scarcity.

This episode is an illustration of a statement made repeatedly in previous chapters namely that speculation is not limited to financial markets but also occurs in other markets for instance in the markets for commodities, real estate or collectibles. Such speculative movements result in price peaks, which share many common characteristics: same order of magnitude of duration with respect to amplitude, similar shape (as we have seen in a previous chapter). However, a closer examination suggests that there are in fact (at least) two distinct classes of speculative peaks, which will be referred to as the U-class and the S-class. From a theoretical point of view this suggests that different models are required in order to describe these distinct kinds of speculative behavior.

The study proceeds as follows: first, in order to provide an intuitive overview, I present speculative peaks for two illustrative cases. Then, by applying two different criteria, I show that these cases belong to two different classes. In a subsequent step I point out that the two criteria are in fact closely connected. Then, in order to get a clearer idea of the dividing line between the two classes an enlarged sample of

157

goods is analyzed. Finally, I point out that the U- and S-classes also differ in terms of response times.

The structural regularities pointed out in the present chapter prepare the way for the theoretical description proposed in the next chapter.

1 Speculative peaks: two illustrative examples

1.1 Wheat price peaks

Historically the most frequent speculative peaks were certainly those for grain prices; if we restrict ourselves to Western Europe it is wheat which played the key role. For centuries wheat was the most traded commodity and at the same time the most important foodstuff; in that sense wheat played a role similar to that of oil nowadays. Fig. 7.1(a) shows two typical price peaks in France at the beginning of the nineteenth century. Why should such price movements be considered as speculative peaks? After all, it is easy to imagine a scenario where a bad harvest (due to unfavorable weather conditions) brings about a price increase without the intervention of speculation. At the start of the episode there is certainly an exogenous shock but this explains only the first phase. As explained at the beginning of this chapter, in specific circumstances the speculative behavior of wheat traders tended to amplify price fluctuations; to see the matter more clearly let us consider a parallel with a twentieth-century situation, namely the case of the oil industry at the beginning of the war between Iraq and Iran (1980–1988). As both countries were major oil producers the war triggered an increase in oil prices because of a drop in supply. But once prices had begun to rise all agents in the market tried to make the best of it, which meant that oil companies, producing countries and industries holding large stocks devised production and inventory strategies with the objective of optimizing the windfall profit generated by the bubble. A similar scenario occurred for wheat markets in past centuries. Indeed one knows both from contemporary accounts and from historical analysis (e.g. Usher 1913, Vuitry 1885) that as soon as a substantial price increase was foreseen farmers, millers, wholesale traders, and even wealthy consumers began to build up stocks. Such a view is confirmed by the fact that the shape of all major price peaks is almost the same and follows the so-called sharp peak–flat trough pattern (for more detail see chapter 6). If the peaks were due to meteorological factors their shapes would be as random as climatic fluctuations; in that respect it should be recalled that precipitations, for instance, are almost as erratic as white noise (see Roehner 2001a, 154).

1.2 Real estate prices

A second example is given in fig. 7.1(b) which describes the movement of real estate prices in the UK during the 1980s. Each curve refers to a specific region;

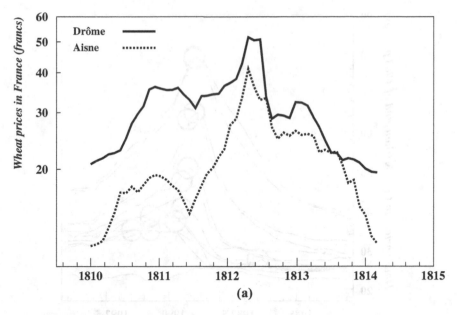

(a)

Fig. 7.1a. Wheat price in France (1810–1814)
Notes: The prices are expressed in franc per hectoliter. Drôme and Aisne are two départements which are about 500 kilometers apart.
Source: Labrousse, Romano, and Dreyfus (1970).

for instance, the two highest refer to the London area and the South-East region while the lowest refers to Northern Ireland (in this case there is in fact no price peak at all). The observation that the curves are smoother than in fig. 7.1(a) reflects the fact that the fluctuations in property prices are much slower than for other speculative prices; this may be a consequence of the long delay required to carry out a real estate transaction.

A similar example is given in fig. 7.1(c) which describes the movement of residential land prices in the 47 prefectures composing Japan during the speculative episode of the late 1980s and early 1990s.

Each of the above graphs shows that the maxima are dispersed over a fairly large time interval. More precisely the regions characterized by the highest prices peak first, while those with lower prices follow the movement with a delay which depends upon proportional to their price level. In other words the speculative bubble originated in the center of the capital city and from there it spread to the other regions. In the UK the speculative wave reached Scotland about two years after London; in Japan it reached Hokkaido about four years after Tokyo. A regression analysis between distances to the capital city and the localizations in time of the different maxima permits an evaluation of the velocity of the price wave; the results are as shown in table 7.1.

It should be noted the wave speeds of the three bubbles are not directly comparable in the sense that they refer to different items.

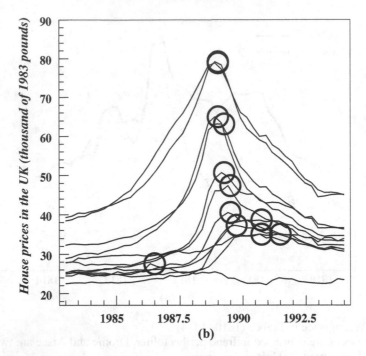

(b)

Fig. 7.1b. House prices in the UK during the speculative episode of 1984–1994
Notes: Each curve represents the price in one of the 12 regions composing the United Kingdom. The two highest curves correspond to "Greater London" and "South East". The circles
indicate the positions of the maxima. Note that the maximum of the lowest curve which
corresponds to Northern Ireland is obviously spurious for this curve has no real maximum.
Source: UK property data are collected and published by the Halifax group (West Yorkshire,
England). I am most grateful to the people at Halifax for their kind assistance.

In the following sections we will see that wheat price peaks and real estate peaks
differ in several ways, a clear indication that their underlying dynamic mechanisms
are certainly fundamentally different.

2 The price multiplier criterion

Let us denote by p_1 the price at the start of the bubble and by p_2 the peak price;
in the following I examine whether there is a relationship between the amplitude
of the peak $A = p_2/p_1$ and the price p_1. More specifically I will compare p_1 and
A for several "realizations" of a given price bubble. What do we mean by different realizations? For wheat prices it means prices on various spatially separated
markets; for stamps it means prices of different stamps; for real estate it means
property prices in different regions. By analyzing a sample of speculative items we
can hope to learn something about the space-time structure of price peaks.

Table 7.1.

Region	Item	Period	Wave speed [km/year]	Correlation
United Kingdom	Houses	1984–1994	190 ± 50	0.93
Japan	Land	1984–1994	72 ± 38	0.50
Paris	Apartments	1986–1995	7.8 ± 2.5	

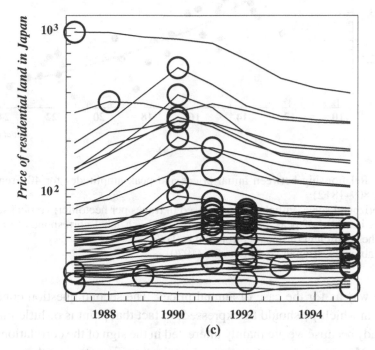

(c)

Fig. 7.1c. Land prices in Japan during the speculative episode of 1986–1995
Notes: Each curve represents the price in one of the 47 prefectures composing Japan. The two curves with the highest maxima correspond to Tokyo and Osaka respectively. For several of the lowest curves there is no clearly defined maximum which means that some of the positions indicated by the circles may be spurious. As can be seen the regions with the highest prices peaked first, while the regions with lower prices followed the movement with a delay inversely proportional to the price level.
Sources: Japan Statistical Yearbook and *Jetro Business Facts and Figures* (various years).

Before coming to the results two technical points must be clarified. The first concerns the very identification of p_1. When one uses annual prices p_1 is simply the first price in a series of steady increases; when one uses monthly prices there are usually some accessory fluctuations. In the latter case one first performs a moving average which smoothes out all minor fluctuations and enables us to define p_1 in

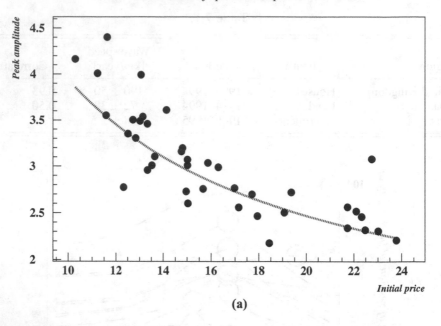

(a)

Fig. 7.2a. Relationship between initial prices and peak amplitudes for 40 French wheat markets (1809–1812)
Notes: Horizontal scale: initial price expressed in francs per hectoliter; vertical scale: peak amplitude, i.e. ratio of peak price to initial price. Each label corresponds to a different market. The coefficient of linear correlation is 0.67.
Source: Labrousse, Romano, and Dreyfus (1970).

the same way as for the case of annual prices. The second question concerns the currency in which p_1 should be expressed; in fact that point is of little importance in this study because we are mainly interested in the sign of the correlation between A and p_1. Moreover when estimating regression coefficients we rather consider the regression coefficient between $\log p_1$ and A and the latter is independent of the unit of currency used.

The evidence about the relationship between p_1 and A for wheat and property markets is summarized in fig. 7.2(a), (b). The first figure corresponds to prices on 40 French wheat markets during the peak which culminated in 1812. The first fact which must be stressed is that there is indeed a definite relationship between initial prices and amplitudes; given the fairly chaotic behavior of commodity prices this comes as a rather good surprise. Secondly, one can observe that there is a *negative* correlation.

The evidence for real estate markets shows that there is also a definite relationship but that, in contrast to wheat markets, there is a *positive* correlation. This means that the amplitude of the peak is larger for expensive areas than for cheap areas. Subsequently this will be referred to as the price multiplier effect. Before trying to

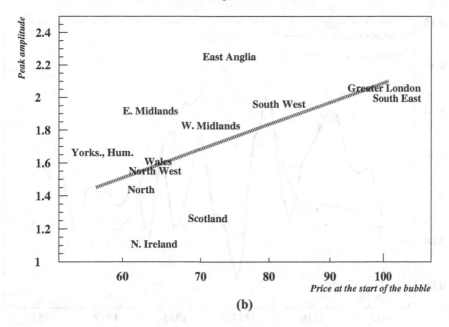

(b)

Fig. 7.2b. Relationship between initial prices and peak amplitudes for property prices in 12 regions of the United Kingdom

Notes: Horizontal scale: prices expressed in thousand euros of January 1, 1999; the vertical scale represents the ratio p_2/p_1 of the peak price to the initial price. The correlation is 0.59 (confidence interval at probability 0.95 is 0.02, 0.87).

Source: Same as for fig. 7.1b.

interpret the difference observed between wheat and property markets it will be of interest to use another criterion.

3 The ensemble dispersion criterion

Given a sample of wheat markets I can at any moment define their spatial mean m and standard deviation σ; for instance if in February 1811 the prices at Paris, Marseilles, and Toulouse were 20, 23, and 26 francs the mean will be 23 and the standard deviation be 3; as a result the coefficient of variation $c = \sigma/m$ will be $3/23 = 13\%$. By repeating that calculation for each month one can study the movement of the spatial coefficient of variation over the course of time. Applying that procedure to the wheat and real estate price series considered in the previous paragraph one obtains the results summarized in fig. 7.3(a), (b). For wheat markets the coefficient of variation shows a trough while for real estate markets it has a peak. In the second case the curve turns out to be much smoother for two reasons: first, we used monthly prices for wheat but quarterly prices in the case of real estate and, second, in a general way property prices are smoother than commodity prices.

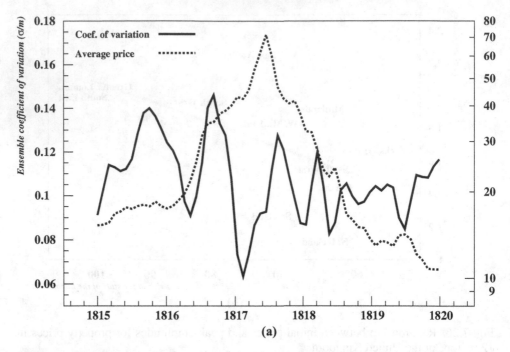

Fig. 7.3a. Ensemble coefficient of variation for 13 wheat markets in Bavaria during the 1815–1819 price peak
Notes: The average price (right-hand scale) is expressed in gulden. The large fluctuations are due: (i) to the fact that we use monthly prices, (ii) to the small number of markets. If one disregards the seasonal fluctuations (peak in harvest time, trough in winter) there is a distinct minimum of the coefficient of variation which precedes the price peak by about four months. The solid line curve has been smoothed through a (centered) moving average with a window of three years.
Source: Seuffert (1857).

The obvious question which presents itself is whether this result is related to the previous one. The answer is affirmative as we will now see. For this purpose one needs the following lemma.

Lemma-conjecture:

X is a random variable with a continuous, unimodal density function.
$g(x)$ is a differentiable, increasing function that is either convex or concave.
$c(X)$ denotes the coefficient of variation of X: $c(X) = \sigma_X / m_X \quad c(X) > 0$.
Then, one has the following inequalities for $c[g(X)]$ (which is supposed to exist):

$\quad g$ convex $(g'' > 0) \Longrightarrow c[g(X)] > c(X)$
$\quad g$ concave $(g'' < 0) \Longrightarrow c[g(X)] < c(X)$

I called that lemma a conjecture because I did not find it in any probability treatise. Without proposing here a mathematical proof, which would be out of order here

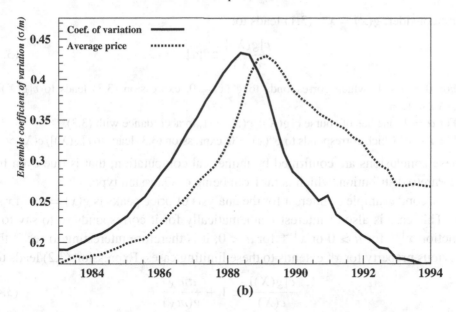

Fig. 7.3b. Ensemble coefficient of variation for house prices in 12 regions of the United Kingdom
Notes: The fact that the curve is less spiky than for wheat prices can be attributed to the fact that house prices are less volatile than wheat prices. The peak in the coefficient of variation occurs about nine months before the peak for the spatial average.
Source: Same as for fig. 7.1b.

present book, I would like to give some plausibility arguments. At the same time this will show the kind of applications that we have in mind.

First we recall a standard approximate estimate of the mean and variance of $g(X)$ (Papoulis 1965, 151). Suppose that, similarly to a Gaussian variable, the random variable X has a density function which takes significant values only in an interval near m_X of the order of its standard deviation, then:

$$E[g(X)] \simeq g(m_X) + g''(m_X)\frac{\sigma_X^2}{2} \quad \sigma_{g(X)} \simeq |g'(m_X)|\sigma_X \qquad (3.1)$$

These expressions should be seen as the beginning of an expansion in powers of σ_X; note that for a Gaussian variable a complete expansion of this kind can be written explicitly. From (3.1) it results that:

$$\frac{c[g(X)]}{c(X)} \simeq \frac{|g'(m_X)|m_X}{g(m_X)} \qquad (3.2)$$

Let us now apply this formula to two specific examples. In each case our conclusions will be checked through numerical computation for three distributions: Gaussian, log-normal, and truncated Pareto with an exponent equal to 1.

First consider: $g(x) = x^\alpha$; (3.1) leads to:

$$\frac{c[g(X)]}{c(X)} \simeq |\alpha| \qquad (3.3)$$

- For $0 < \alpha < 1$, which corresponds to $g''(x) < 0$, expression (3.3) leads to $c[g(X)]/c(X) < 1$.
- For $\alpha = 1$, one has of course $c[g(X)]/c(X) = 1$, in accordance with (3.3).
- For $1 < \alpha$, which corresponds to $g''(x) > 0$, expression (3.3) leads to $c[g(X)]/c(X) > 1$.

These conclusions are confirmed by numerical computation; that is true even for the Pareto distribution, which is far from being of Gaussian type.

   A second example of interest for the analysis of price peaks is $g(x) = ax \ln x + bx$. This case is also of interest mathematically for it corresponds so to say to a function $x^{1+\epsilon}$ for $a > 0$ or $x^{1-\epsilon}$ for $a < 0$; it is therefore interesting to see if the previous property for x^α extends to these limiting cases. Expression (3.2) leads to:

$$\frac{c[g(X)]}{c(X)} \simeq 1 + \frac{am_X}{g(m_X)} \qquad (3.4)$$

In subsequent applications, $g(m_X)$ represents a price and is therefore positive; moreover, the second derivative of $g(x)$ is $g''(x) = a/x$. Thus:

- If $a < 0$, which corresponds to $g''(x) < 0$, $c[g(X)]/c(X) < 1$.
- If $a > 0$, which corresponds to $g''(x) > 0$, $c[g(X)]/c(X) > 1$.

Let us now apply the above lemma to wheat and real estate prices.

Wheat prices

$$p_2/p_1 = -ap_1 + b \quad (a > 0) \Longrightarrow p_2 = -ap_1^2 + bp_1 \Longrightarrow p_2'' = -2a < 0$$

thus the function is concave and $c(p_2) < c(p_1)$ as we indeed observed.

Real estate prices

$$p_2/p_1 = ap_1 + b \quad (a > 0) \Longrightarrow p_2 = ap_1^2 + bp_1 \Longrightarrow p_2'' = 2a > 0$$

thus, the function is convex and $c(p_2) > c(p_1)$ as we indeed observed.

The same conclusions apply if one uses (as is done in table 7.1) the adjustment $p_2/p_1 = a \ln p_1 + b$. The motivation for replacing p_1 by $\ln p_1$ is to make the values of a independent of the units of currency and therefore to make them comparable for different cases; needless to say, the two adjustments lead to the same results as far as the sign of a (which is our main interest in this paper) is concerned.

   A last remark is in order. A relationship of the form $Y = g(X)$ would correspond to a correlation between X and Y which is equal to 1. For real data (see the previous section) the correlation is of course always smaller than 1; this would therefore correspond to a relationship of the form $Y = g(X) + \epsilon$, where ϵ represents a random noise of mean 0 and standard deviation σ_ϵ. If, as is natural, one assumes that ϵ is independent of X, Y only one factor, namely σ_Y, will be modified in the ratio $c(Y)/c(X)$; and since:

$$\sigma_Y^2 = \sigma_{g(X)}^2 + \sigma_\epsilon^2$$

one comes to the conclusion that the ratio $c(Y)/c(X)$ will increase. The result stated in the lemma will therefore be even better observed when $g'' > 0$; however in the case of $g'' < 0$, if σ_ϵ is large enough with respect to $\sigma_{g(X)}$ the result given in the lemma may no longer hold. This will indeed be observed in some of the cases considered in the next section.

In short, the second criterion is in fact a consequence of the first. The previous criteria define two types of speculative peaks; naturally, the question which presents itself is whether other goods follow the same classification scheme.

4 Two classes

In this section we broaden our perspective to a wider sample of speculative markets. In all the graphs of fig. 7.4 and 7.5 the solid and dashed lines respectively correspond to the highest and lowest initial prices. Since all initial prices have been normalized to 100 the amplitudes are given by the ordinates of the peaks. Figs. 7.4 summarize some evidence for wheat and potato markets in Europe and in the United States, both in the nineteenth and twentieth centuries. For all graphs the amplitude is largest for the smallest initial price, which means that the effect previously observed for wheat markets was not specific to nineteenth-century European markets.

The reader may wonder why we systematically consider regional instead of price series for different countries. The reason is simple: in the latter case the comparison is usually biased by the incidence of tariff barriers or exchange rates. Another question is why wheat and potato markets are of particular interest. Wheat and potatoes are two very different commodities: wheat is extensively traded on international markets to the point that it is often said that wheat prices in the United States are driven by exports; in contrast international trade is of marginal importance for potatoes. Yet, in spite of such differences, the spatial pattern is the same for both commodities.

Figs. 7.5 summarize some evidence for diamonds, American collectible coins, and French and British postage stamps. In all these cases the amplitude is larger for the highest initial price (solid line).

Of course the evidence provided in figs. 7.4 and 7.5 for pairs of markets must now be extended to larger sets of markets; this is done in table 7.2. The markets

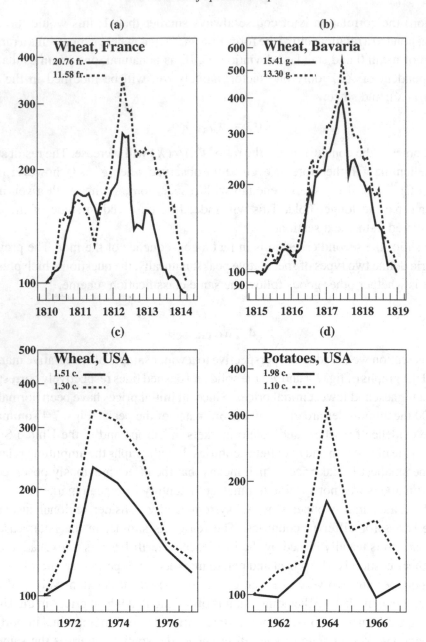

Fig. 7.4a,b,c,d. Price peaks for pairs of items differing by their initial price (U-class)
Notes: For all the panels in figs. 7.4 and 7.5 the solid line corresponds to the largest initial price while the dashed line corresponds to the smallest initial price. (a) Wheat prices in France. Solid line: Drôme (south of France); dashed line: Aisne (north-east); monthly prices. (b) Wheat prices in Germany. Solid line: Bayreuth; dashed line: Nurnberg; monthly prices. (c) Wheat prices in the United States. Solid line: North Dakota; dashed line: Nevada; annual prices. (d) Potato prices in the United States. Solid line: Rhode Island; dashed line: Minnesota; annual prices.
Sources: Labrousse, Romano, and Dreyfus (1970), Seuffert (1853), Langley and Langley (1989), Lucier *et al.* (1991).

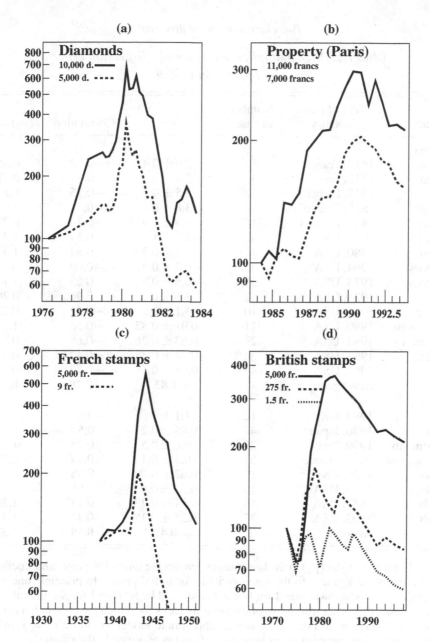

Fig. 7.5a,b,c,d. Price peaks for pairs of items differing by their initial price (S-class)
Notes: (a) Prices of polished diamonds on the Antwerp market. Solid line: one carat and D-clarity; dashed line: one carat and G-clarity. (b) Property values in Paris. Solid line: price of apartments in one of the most expensive arrondissements (7th); dashed line: price of apartments in one of the less-expensive arrondissements (19th) (c) Prices of French postage stamps. Solid line: one of the most expensive French stamps ("Yvert et Tellier" number 2, not postmarked); dashed line: price of a fairly cheap stamp ("Yvert et Tellier" number 16, postmarked). (d) Prices of British postage stamps. Solid line: one of the most expensive British stamps ("Yvert et Tellier" number 90); dashed line: medium-price stamp ("Yvert et Tellier" number 286); dotted line: price of a fairly cheap stamp ("Yvert et Tellier" number 106).
Sources: Diamonds: Diamonds 1988, *The Economist*, Special Report No. 1126; Property: *Chambre des Notaires* (Paris); French stamps: Massacrier (1978); British stamps: "Yvert et Tellier" catalogs, various years.

Table 7.2. *Price multiplier and ensemble dispersion criteria*
$$p_2/p_1 = a \ln p_1 + b$$

Product	Year of peak, country	Number of cases	a	Correlation	$c(p_2)/c(p_1)$
U-class					
Wheat	1812, France	40	-2.0 ± 0.4	-0.82	0.56
Wheat	1812, France	40	-2.3 ± 0.5	-0.84	0.47
Wheat	1817, France	40	-2.5 ± 0.7	-0.75	0.83
Wheat	1817, France	40	-3.8 ± 0.7	-0.73	1.08
Wheat	1817, Bavaria	12	-3.8 ± 1.9	-0.79	0.70
Wheat	1973, USA	42	-2.8 ± 1.3	-0.54	1.35
Wheat	1980, USA	42	-1.3 ± 0.3	-0.84	1.35
Potatoes	1964, USA	35	-0.7 ± 0.3	-0.69	0.73
Potatoes	1973, USA	35	-0.6 ± 0.8	-0.25	1.14
Cotton	1934, USA	10	-1.6 ± 0.2	-0.98	0.26
Cotton	1946, USA	10	-3.1 ± 0.5	-0.97	0.27
Sugar beets	1990, USA	11	-0.91 ± 0.83	-0.58	1.18
Soybeans	1988, USA	29	-0.63 ± 0.26	-0.68	0.74
Apples	1991, USA	36	-0.74 ± 0.38	-0.42	0.97
Corn	1988, USA	41	-0.65 ± 0.19	-0.73	0.70
	Average		**-1.83**	**-0.70**	**0.82**
S-class					
Houses	1989, UK	12	1.1 ± 1	0.59	2.20
Land	1990, Japan	47	0.45 ± 0.2	0.59	1.21
Apartments	1990, Paris	20	0.6 ± 0.5	0.48	1.40
Post. stamps	1944, France	57	0.3 ± 0.1	0.49	0.83
Books	1929, USA	10	0.62 ± 0.09	0.98	
Books	1929, USA	10	0.33 ± 0.05	0.96	
Bonds	1857, USA	13	0.11 ± 0.43	0.13	1.30
Bonds	1893, USA	37	0.25 ± 0.18	0.42	1.31
	Average		**0.49**	**0.59**	**1.37**

Notes: The ratio $c(p_2)/c(p_1)$ in the last column denotes the ratio of the ensemble coefficient of variation of peak prices to the same variable for initial prices. In principle one would expect this ratio to be lower than 1 for U-class cases, and larger than 1 for S-class cases, but, in a few cases, this is not true. Most often these are cases for which the correlation between p_1 and p_2 is fairly poor. For these cases an appropriate model would be $p_2 = f(p_1) + \epsilon$ (where ϵ is a noise variable) rather than $p_2 = f(p_1)$ as assumed in the lemma.

Sources: Wheat, France: Labrousse, Romano, and Dreyfus (1970); wheat, Bavaria: Seuffert (1857); wheat, USA: Langley and Langley (1989); potatoes: Lucier *et al.* (1991); houses: Halifax group; land: *Japan Statistical Yearbook, Jetro Business Facts and Figures* (various years); postage stamps: Massacrier (1978); books: *American Book-Prices Current* (various years); bonds: Macaulay (1938).

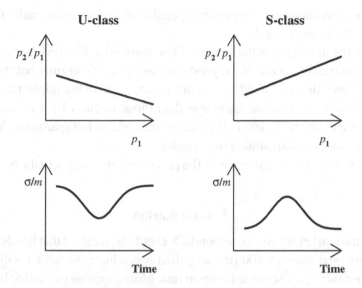

Fig. 7.6. Two classes of speculative peaks
Notes: The figure schematizes two characteristics by which the two classes differ: (i) peak amplitude with respect to initial price level, (ii) movement of the coefficient of variation in the course of time.

which follow the wheat price pattern are referred to as U-class markets; the markets which follow the real estate price pattern are referred to as S-class markets. That distinction is summarized in fig. 7.6.

What are the distinctive features of S-class goods with respect to U-class goods? One fundamental characteristic is the fact that for S-class goods investors can select one item among a number of similar ones at a small cost. For instance a stamp collector can choose to buy stamp *A* rather than stamp *B* at no additional cost apart from the time it takes to make the selection. At first sight it could seem that houses and apartments are a perfect counter-example: if I want to live in Scotland I cannot buy a house in London. One must admit that property bubbles are mainly engineered by investors, that is to say people who buy to re-sell with a profit, rather than by users; clearly for investors it does not matter whether they buy in Scotland or in London. In contrast for a wheat trader there is a big difference between buying in Kansas or Oregon: in the first case the wheat has to be shipped to the East or West Coast before it can be exported; in the second case it can be directly exported through the harbor of Portland in northern Oregon. The same observation can be made for the other goods in the U-class: in each case the location where the good is produced and sold really matters.

There are of course other differences between U- and S-class markets. For instance S-class goods can be easily stored even for a long time, whereas U-class goods can hardly be stored for more than a few months and their storage is costly.

This feature also contributes to making speculative trading much easier for S-class goods than for U-class goods.

Other distinctive features that one could be tempted to mention are largely irrelevant: (i) It is true that most of the goods in class U are foodstuffs but the example of cotton shows that this condition is not essential. (ii) One might think that the goods in class S are less indispensable than those in class U; it is true that collectible postage stamps are less indispensable than wheat, but apartments and houses are certainly more indispensable than apples.

In the next section we examine how the previous results apply to the bond market.

5 Bond market

I consider the market for corporate bonds. It should be recalled that bonds are issued at a conventional price of 100 (the so-called par-value); the holder will receive a fixed interest rate r, the so-called coupon rate. Moreover, one can sell or buy a given bond at any time; thus, as for a share, its price will rise or decrease depending on the potential number of sellers and buyers. Suppose that two years after it has been issued the price of a bond has decreased to 90; the new buyers will receive the same coupon rate, but since they bought the bond for less than 100 their real interest rate will be $r/0.9 = 1.1r$, a quantity which is called the current yield of the bond. Viewed from the side of the company which issued the bond, the yield is the price paid in order to get additional capital. During nineteenth-century stock market contractions, bond yields usually exhibited a peak, because these crisis were often marked by a credit crunch. Since there are different bonds among which investors can make a selection, one might posit that bond yields belong to the S-class. Let us see if this is true.

As a case in point I consider the crash from January to October 1857. The average yield rose from about 8 to 10 percent (fig. 7.7(a)), while the coefficient of variation for 13 different bonds rose from 13 to 17 percent; the latter observation is a first indication that bonds do indeed belong to the S-class. For the correlation between initial yields and amplitudes (i.e. ratios between peak yields and initial yields) one gets a poor 0.13 (confidence interval at probability level 0.95 is −0.45 to 0.63); however fig. 7.7(b) shows that this low correlation level is mainly due to the occurrence of three outliers; without these outliers the correlation would be markedly larger.

Remark. It can be objected that the present investigation does not really parallel previous ones in the sense that we consider yields rather than prices; however it can be argued that when buying bonds investors are attentive to the interest rate

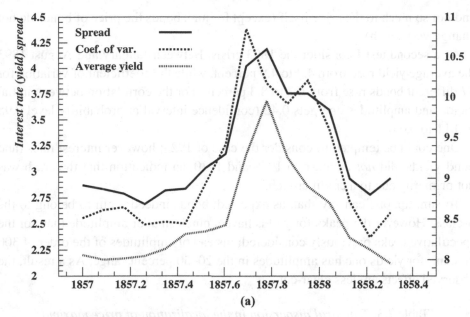

Fig. 7.7a. Yield of corporate bonds during the crash of 1857
Notes: In addition to the coefficient of variation we have also shown the spread, which is a variable commonly used in bond analysis; here it is defined as the difference between the 25 percent highest yields and the 25 percent lowest yields; as can be seen it follows closely the coefficient of variation.
Source: Macaulay (1938).

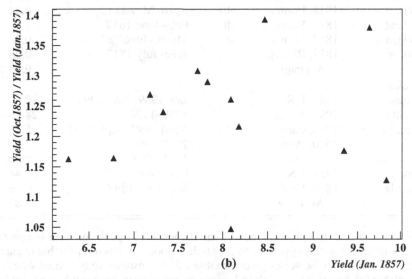

Fig. 7.7b. Relationship between initial yields and the ratio of peak yields to initial yields
Notes: The correlation is low (0.13) and non-significant but this is mainly due to the three outliers in the lower-right corner of the figure. By and large the properties of corporate bonds resemble those of S-goods.
Source: Macaulay (1938).

and not so much to the price itself (except for junk bonds the price of bonds do not change very much).

As a second test I consider the 1893 crisis. Between January and August 1893 the average yield rose from 4.4 to 4.9 percent, while the coefficient of variation for 37 different bonds rose from 6.3 to 8.3 percent. For the correlation between initial yields and amplitudes one gets 0.42 (confidence interval at probability level 0.95 is 0.11 to 0.65).

One could be tempted to consider the case of 1929; however interest rates (and bond yields) did *not* increase in 1929 and 1930, an indication that the crash was not primarily due to a credit crunch.

To sum up, one can say that, as expected, bonds indeed seem to belong to the S-class. However the peaks for yields have a much smaller amplitude than for the speculative peaks previously considered; instead of amplitudes of the order of 300 percent, for yields one has amplitudes in the 20–30 percent range. As a result, the observed effects are less clear-cut.

Table 7.3. *Temporal dispersion in the localization of price maxima*

Product	Year of peak, country	Number of cases	Earliest and latest date	Standard deviation [month]
U-class				
Wheat	1812, France	40	March–July 1812	0.74
Wheat	1812, France	40	April–May 1812	0.41
Wheat	1817, France	40	Feb.–June 1817	1.1
Wheat	1817, France	40	March–June 1817	0.98
Wheat	1817, Bavaria	12	June–July 1817	0.28
	Average			**0.70**
S-class				
Houses	1989, UK	12	Jan. 1989–June 1991	10.3
Land	1990, Japan	47	1989–1994	24.1
Apartments	1990, Paris	20	April 1990–Sept. 1991	5.36
Offices	1990, Paris	2	1990–1991	12.0
Post. stamps	1944, France	57	1941–1946	9.26
Bonds	1857, USA	13	Oct.–Nov. 1857	0.44
Bonds	1893, USA	37	July–Oct. 1893	0.66
	Average			**8.87**

Notes: With the exception of bonds, the dispersion in peak maxima is several times larger for S-class goods than for U-class goods. It was not possible to carry out that test for all the series given in table 7.2; for instance only annual data are available for American agricultural prices by state, whereas monthly (or weekly) prices would be required. The two office items are new and old offices (the new offices peak first).

Sources: Offices: *Les Echos* (June 27, 1996); for the other goods see table 7.2.

6 Differences in response times

In some cases it is possible to track the diffusion of a speculative bubble and even to estimate the diffusion velocity. Real estate bubbles are particularly convenient in that respect because of their long characteristic time. At the beginning of this chapter we gave some wave speed estimates (for more details see Roehner 1999a and b, 2001a).

In many cases it is not possible to estimate the wave velocity of speculative bubbles, either because the high-frequency data that would be required are not available or because one cannot define a suitable notion of distance: how, for instance, can one define a distance in the space of postage stamps? However it is possible to characterize the response time of different markets by considering the two following features: (i) the dispersion of peak times and (ii) the relationship between peak amplitudes and peak times. These are what we examine now.

6.1 Dispersion of peak times

During a speculative episode the prices on various wheat markets reach their maxima within one or two months. On the contrary property prices in different regions

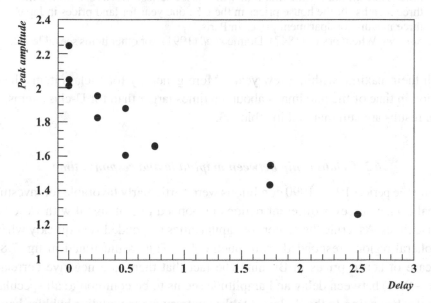

Fig. 7.8. Relationship between amplitude and delay for house prices in Great Britain
Notes: Horizontal scale: delay (expressed in years) with respect to the earliest peak; vertical scale: amplitude of the peak (that is ratio of peak price to initial price). The data are quarterly prices for the period 1984–1994 in the 11 regions composing Britain. The coefficient of linear correlation is −0.91; however it is fairly clear that the data points exhibit a non-linear trend; this will be further explained in the next chapter.
Sources: See Fig. 7.1b.

Table 7.4. *Relationship between amplitude and response time*

Product	Year of peak, country	Number of cases	Correlation
U-class			
Wheat	1812, France	40	0.07
Wheat	1812, France	40	0.10
Wheat	1817, France	40	0.34
Wheat	1817, France	40	0.44
Wheat	1847, France	40	−0.023
	Average		**0.18**
S-class			
Houses	1989, UK	12	−0.91
Land	1990, Japan	40	−0.46
Apartments	1990, Paris	20	−0.23
	Average		**−0.53**

Notes: The response time is defined as the delay with respect to the earliest peak. For U-class items the correlation is either positive or negligible; for S-class items it is negative. This test requires data whose sampling time is small with respect to the dispersion of peak times (estimated in table 7.3); the sampling time is one month for the first four wheat price series; two weeks for the fifth; three months for the house prices in the UK; one year for land prices in Japan; three months for apartment prices in Paris.
Sources: Wheat prices (1847): Drame *et al.* (1991); for other items see table 7.2.

reach their maxima within a few years. More generally for S-class items the dispersion in time of the maxima is about ten times larger than for U-class items. The main results are summarized in table 7.3.

6.2 *Relationship between amplitude and response time*

During the period 1985–1990 conditions were particularly favorable for investment in real estate; however different regions responded to that signal with various response times. As a rule the centers of capital cities responded very quickly whereas peripheral regions responded with much delay. This is illustrated in fig. 7.8 for the case of house prices in Britain. The fact that there is a negative correlation ($r = -0.91$) between delay and amplitude seems to be common to all speculative episodes belonging to the S-class. On the contrary for speculative bubbles belonging to the U-class the correlation is either positive or close to 0 (that is to say, not significant). Several comparative results are summarized in table 7.4

8

Dynamics of speculative peaks: theoretical framework

Between 1938 and mid 1943 the price of nineteenth-century French stamps was on average multiplied by a factor three. In a previous chapter we interpreted that movement as a hedge against inflation. Such an interpretation is indeed confirmed by the fact that other inflationary periods were also marked by substantial peaks for stamp prices. Yet, if inflation can with good reason be considered as a triggering factor, the evidence suggests that price peaks have their own dynamics: once a speculative movement has been started it develops on its own momentum and ends when that momentum is exhausted. This is precisely what happened in 1943. Between 1939 and 1943 the average annual inflation rate in France was 18 percent; in combination with one or several other factors, which we do not yet understand, that inflation activated the speculation. The conclusion that the movement had its own dynamics is obvious from the fact that the turning point occurred in mid 1943, in spite of the fact that the inflation rate was higher in 1943 than in any previous year. After the downturn stamp prices continued to fall in spite of the fact that during the five following years the average annual inflation rate was as high as 46 percent. Somehow the momentum of the bubble was exhausted and even a huge inflation rate was unable to trigger another bubble. The same kind of phenomenon occurred for the gold and silver bubble in January 1980: the bubble was triggered by double-digit inflation rates, but it burst in spite of the fact that in 1980 the inflation rate in the United States reached the record level of 14 percent (the highest figure since 1974).

These examples illustrate one of the main ideas of this chapter: because of their own dynamic structure speculative markets may react to activating factors with a delay of several years and, once the downward phase has begun, even a reinforcement of the exogenous factor is unable to stop the fall. A similar observation can be made also for stock markets. Thus, after the crash of October 1929 the price of stocks continued to fall, in spite of a massive reduction in discount rates from 6 percent to a level of 2 percent within 14 months.

177

It is deliberate that in the title of this chapter we used the term "theoretical frame-work" rather than "model." Indeed our objective is not to build a model for a specific market but to present a theoretical perspective from which several properties of speculative peaks may be interpreted. In so doing we follow a standard physical approach; for instance, once the concept of damped vibrations has been understood and formalized, it may be applied to several physical systems, e.g. electromagnetic damping of a galvanometer, damping in plasma vibrations, or collision damping in the ionosphere (in relation to the propagation of radio waves). Such an approach has both advantages and drawbacks. One of the main advantages is that it may explain and unify a broad array of facts previously unconnected; subsequently such a framework can be progressively improved and refined. In contrast, a specialized model often is merely superseded by another, which is based on alternative assumptions and may account for the same (too narrow) set of facts. The main drawback of the framework approach is that, being a loose patchwork rather than a full-fledged theory, it may at first appear fairly unconvincing.

In this chapter we interpret many of the regularities observed previously, in the light of a simple dynamic structure; even in its non-stochastic (i.e. deterministic) version this structure is able to account for several of the phenomena that we want to describe. In fact the difficult part of the task is not to solve the equations but to select, collect, simplify and arrange the regularities on which they are based.

1 Main ideas

In what follows we will describe speculative peaks as the response of a dynamic system to an exogenous shock. If we denote by H the differential operator which describes the system, its impulse response $G(t)$ to a shock that occurred at the time $t = 0$ will be defined by:

$$HG(t) = \delta(t) \tag{1.1}$$

where $\delta(t)$ denotes the Dirac distribution. Although such a description is standard in physics or electrical engineering, its application to economic problems rests on a number of unconventional concepts which are discussed in the following paragraphs. In order to make the discussion more concrete let us consider two specific examples: (i) the price peaks that occurred at the beginning of 1980 for cobalt, diamonds, gold, palladium, platinum, and silver (see fig. 1.3(a)) and (ii) the price peaks that occurred in real estate markets in 1985–1995 (see fig. 7.1(b)).

1.1 A comparative perspective

First it must be emphasized that in contrast to most econometric models we shall not describe the markets for diamonds, gold, or silver as separate markets. This makes sense because we restrict ourselves to speculative bubbles during which the price movements on these markets are closely connected. In more normal times these markets have only a loose connection and a more detailed description would be required. This situation is similar to what happens in physical systems during phase transitions. As one knows, when close to a critical transition point all systems have similar behavior irrespective of their detailed structure. In short, it is the fact that we focus on speculative episodes which enables us to resort to a simple description.

1.2 Shock versus permanent monitoring

Considering a speculative peak as the response to a *single* shock may at first sight appear as an unrealistic simplification. The conventional view is rather that one or several factors create a climate which is favorable to the development of speculation and the bubble lasts for so long as these factors are present. However that view is not consistent with the evidence; as we have already pointed out at the beginning of this chapter, the favoring factor may still be present when the turning point occurs.

As another example one may consider the real estate bubbles of the period 1985–1995. If one considers a single country, Japan for instance, it is of course easy to enumerate plausible reasons (the special bond between the Japanese and land, the mochiai system; see Taniguchi 1993) which may "explain" the Japanese speculative bubble. But it is more difficult to explain why there were at the same time similar bubbles in at least half a dozen regions (Australia, Britain, California, France, Sweden, etc.) whose demography and banking traditions were very different. Moreover it would be even more tricky to explain why the factors (tax exemptions and tax cuts, low real interest rates, etc.), which had favored property speculation for over five years, suddenly foundered within a few months simultaneously in all countries. However, if one accepts the view that these bubbles were the responses to an initial shock, of systems having similar dynamical properties it becomes clear why these bubbles more or less had the same duration and peaked almost simultaneously. At this point we will not try to identify precisely the nature of the initial shock, we simply posit its existence. It may be a combination of factors which brought all these systems into resonance or it may be that the process began in one country and then spread to the others.

1.3 Users and speculators

In previous chapters we have seen that the eleven regions composing Britain responded in very different ways to the speculative bubble of the the late 1980s: the areas around London where prices were already high responded strongly, while the northern regions where prices were low responded very weakly. This led us to assume that each region comprised two kinds of buyers:

- users who buy houses and apartments for personal use;
- investors, speculators, and property developers who make money by buying and selling property values.

Going back to Nerlove (1958) there is in economics a long tradition of models which involve a mix of agents with different dynamic responses. A more recent illustration in the field of stock price analysis is the seminal paper by Lux and Marchese (1999) and more generally all multi-agent models. We will here use that idea in order to model different levels of responsiveness to speculative contagion.

By assuming a proportion of users and speculators that change from one region to another along with the existing price level, one is able to explain that the amplitudes of the peaks follow the price multiplier effect and that prices peaked later in the north of Britain than in the London region. As reasonable orders of magnitude one may for instance assume a proportion of about 45 percent users (and 55 percent speculators) in the London region as opposed to 95 percent users in Scotland.

Mathematically the above assumption can be implemented as follows. We denote by $S^{(1)}$, $D^{(1)}$ the supply and demand function of the users and by $S^{(2)}$, $D^{(2)}$ the same functions for the investors. If k denotes the proportion of investors the total supply and demand functions are:

$$S = (1 - k)S^{(1)} + kS^{(2)} \qquad D = (1 - k)D^{(1)} + kD^{(2)}$$

The equilibrium condition would be expressed by writing that the total excess-supply function $s = S - D$ is equal to 0. Here, we are of course not interested in the equilibrium equation but in the transient price fluctuation resulting from an exogenous shock, namely:

$$s = \delta(t)$$

Thus, if we denote by $s^{(1)}$ and $s^{(2)}$ the excess-supply functions of users and investors the equation describing the dynamics of the price peaks would be:

$$(1 - k)s^{(1)} + ks^{(2)} = \delta(t) \tag{1.2}$$

Such a description was illustrated for real estate speculation in a former paper (Roehner 1999a) where explicit expressions were proposed for the excess-supply functions $s^{(1)}$ and $s^{(2)}$. In the present chapter we will focus on dynamical properties without necessarily trying to relate them to the behavior of individual agents.

1.4 Transaction friction

The last idea is that the responsiveness to speculative contagion is largely determined by the amount of transaction friction. What do we mean by that term?

A zero friction transaction would involve no cost and would be carried out instantaneously. Stock market transactions are the closest approximation to a zero friction transaction: the commission is about 0.02 percent (or less for large transactions) and the delay is of the order of a few minutes. At the other end of the scale we find real estate transactions. In France transaction costs (including taxes) are of the order of 10 percent, that is to say 500 times larger than for stocks. Seen from the buyer's side the transaction delay is about two months, which represents the interval between the moment when an agreement has been reached and the moment when the sale becomes effective. On the seller's side the delay is even longer: results from statistics published by a major realtor (namely "Century 21") show that in 1996 a seller had on average to wait about three months before finding a buyer, giving a total transaction delay of five months. Thus for buyers and sellers together the average transaction delay was $(2 + 5)/2 = 3.5$ months, i.e. about 15,000 times more than for stocks. From these orders of magnitude it is clear that the dynamical response of property markets to an exogenous shock will be considerably more damped than the response of stock markets. This point is developed in more detail in the second section.

It must be observed that the previous estimates of transaction costs leave aside the bid–ask differential. This factor was purposely neglected because most agents do not seem to consider it as a real transaction cost. That is quite understandable for nobody (except perhaps day traders) buys an item just to sell immediately; therefore most people are not aware of the real incidence of the bid–ask margin. For a number of collectibles the commission is small but the bid–ask margin is large. For instance a valuable stamp which was bought from a trader at a price of 500 euros will probably not fetch more than 300 euros if one wants to sell it to another trader, a difference which represents a huge transaction cost of 40 percent. For diamonds the bid–ask differential can reach about 70 percent (Duthy 1978). Ultimately the bid–ask margin depends upon the profit margin expected by the trader, and that margin is naturally higher for markets with a small trading volume than for highly liquid markets, such as equity markets.

One may wonder if there is a relationship between transaction costs and transaction frequency. In the ascending phase of a price peak the answer would in principle be affirmative. Indeed if after having bought an item a trader wants to make a profit he must wait until the price increase exceeds the (total) transaction cost t. Thus, if one can assume that in the ascending phase the price climbs on average at a rate α, the delay between two successive transactions must be longer than $d = t/\alpha$. In other words, high transaction costs tend to impose a slower transaction frequency (a more detailed discussion can be found in Roehner 2001a).

1.5 Agents and markets form a compound

The dynamic properties of a speculative market depend both on its organization and on the behavior of the agents. On stock markets for instance day traders and market makers have a very short response time (of the order of several minutes or less), whereas mutual fund shareholders have a response time of several months (see chapter 5 in this respect). Between these extremes there is of course a wide spectrum of agents with response times ranging from a fraction of a day to several weeks. Naturally, at the level of price fluctuations one can only observe the combined effects of the actions of all these agents (only frequency analysis would permit to distinguish between these various contributions).

Throughout this book we have often emphasized that property markets are simpler than equity markets. This is of course obvious as far as technical sophistication is concerned, but in the light of the previous discussion we see that there is also a purely dynamic reason. Indeed, in property markets, even if investors buy and sell more frequently than users (i.e. residents), the frequency of their transactions is limited by the long transaction delays that are inherent to property markets. In other words, whereas in equity markets one can find agents with a broad spectrum of response times, in real estate markets the response time spectrum is much more narrow. It is therefore not surprising that real estate price fluctuations are smoother than stock price fluctuations.

In previous chapters we have seen that for property markets price peaks are both of smaller amplitude and flatter than for stock markets. We have also observed that whenever there has been in the same period a price peak for stocks and property values, stock prices peaked first and decreased faster than property prices (several examples are given in the previous chapter). This is indeed what would be expected for a system with a shorter response time (see below).

Naturally this is a fairly unrefined test in the sense that it supposes a common exogenous shock which is an assumption which is not obvious for two markets so different as stocks and real estate.

So far our arguments were mainly qualitative (they are summarized in table 8.1), we now try to express them in a more precise mathematical form.

Table 8.1. *Theoretical framework for the description of speculative peaks*

1	Scope	Not a single market, but a collection of markets
2	Exogenous forces	Not continuous monitoring, but a single shock
3	Users and speculators	Small (vs. sizeable) proportion of speculators for cheap (vs. expensive) items
4	Transaction friction	Is for instance 1,000 times smaller for stocks than for property values

Table 8.2. *Recapitulation of empirical regularities*

Regularities and examples	Chapter
1 Correlation between prices of different items is increased during speculative bubbles	1
2 Amplitude (A) and shape of peaks (α) • A(stock prices) $\sim$ 3A(property prices) • α (stock prices) $\sim 0.5\alpha$ (property prices) ○ A(fastly reacting syst.) $>$ A(slowly reacting syst.) ○ α (fastly reacting syst.) $<$ α (slowly reacting syst.)	6
3 Time of occurrence of peaks • Stock prices peak before property prices • Property prices in capital cities peak before those in provinces • Dispersion of peak times larger for S-class items than for U-class goods	5,7
4 Peak amplitude as a function of initial price • Negative correlation for U-class goods • Positive correlation for S-class items	7
5 Volatility during price peak • Trough of coefficient of variation for U-class goods • Peak of coefficient of variation for S-class items	7
6 Distribution of price changes • Decrease of the width when response times of the system increase	5,9

Notes: By the term "system" we understand the system formed by the agents and the specific market (with its organizational rules) in which they invest.

2 Implementation

In this section we analyze a number of implications of the ideas described above. First of all, for the sake of clarity, we list some of the regularities examined in previous chapters that will find an interpretation in the present theoretical framework.

2.1 Recapitulation of empirical regularities

Table 8.2 gives an overview of some of the regularities that we will try to explain. Except for the last one, all these effects refer to the behavior of prices during speculative peaks.

2.2 Dynamic equations: first order

As we already stated we will not try to derive the dynamic equations from micro-economic assumptions about the behavior of agents. At this point what really matters are the dynamic properties of the global system.

The simplest equation of the form (1.1) is:

$$\frac{dG}{dt} + aG(t) = \delta(t) \tag{2.1}$$

Assuming the causality condition $G(t < 0) = 0$ to hold, the solution reads:

$$G(t) = e^{-at} Y(t) \tag{2.2}$$

where $Y(t)$ is equal to 0 for $t \leqslant 0$ and to 1 for $t > 0$.

Because $G_{min} = G(0) = 0$ we cannot define the amplitude of the peak p_{max}/p_{min}. To remedy this problem we define the price as:

$$p(t) = \exp[G(t)] \tag{2.3}$$

Subsequently, when discussing the stochastic properties of prices we will see that another good reason for adopting the definition (2.3) is the fact that the distribution of price changes can be fairly well approximated by a log-normal distribution; this will be precisely the case through (2.3) if we assume (as is natural) the noise to be Gaussian.

For the prices defined by (2.3) the amplitude of the peak is given by:

$$A = \exp[G(0^+)] / \exp[G(0^-)] = e$$

In other words all the peaks have the same amplitude no matter what values the parameter a takes. However, a defines the characteristic time of the system $T = 1/a$. This result is therefore in contradiction with condition 2 in table 8.2. Needless to say, the shape of the peak described by the function (8.2), with its vertical upgoing path, is also at variance with empirical observation.[1] This leads us to look for a second-order equation.

2.3 Dynamic equations: second order

After (2.1) the simplest equation of the form (1.1) is:

$$m\ddot{G} + 2b\dot{G} + cG = \delta(t) \tag{2.4}$$

If (2.4) is interpreted as the equation of a damped oscillator, the physical interpretation of m would be the mass, b would be the friction parameter, and c would be the strength of the force that brings the system back to equilibrium. The solution of (2.4) reads:

$$G(t) = \frac{Y(t)}{m(r_1 - r_2)} \left[e^{r_1 t} - e^{r_2 t} \right] \quad r_1 \neq r_2 \tag{2.5}$$

---

[1] However that second reason is less compelling than the first for it is not our primary objective to describe exactly the shape of the peaks.

where r_1, r_2 denote the roots of the characteristic equation:

$$r^2 + 2\gamma r + \omega_0^2 = 0 \qquad \gamma = b/m, \quad \omega_0^2 = c/m$$

If r_1, r_2 are complex the expression (2.5) becomes (Byron and Fuller 1970, 408):

$$G(t) = e^{-\gamma t} \frac{\sin \omega t}{m\omega} Y(t) \qquad (2.6)$$

where $\omega = \sqrt{\omega_0^2 - \gamma^2} = \sqrt{c/m - (b/m)^2}$.

By going to the limit $\omega \to 0$ in (2.6) one obtains the solution for the case $r_1 = r_2$, namely:

$$G(t) = \frac{t}{m} e^{-\gamma t} Y(t) \qquad (2.7)$$

Fig. 8.1a shows a number of Green's functions for various parameters.

A simple and useful observation can be made regarding the derivative at $t = 0$. If one integrates equation (2.4) over an interval $(-\epsilon, \epsilon)$ one gets:

$$m[\dot{G}(\epsilon) - \dot{G}(-\epsilon)] + 2b[G(\epsilon) - G(-\epsilon)] + c(2\epsilon)G(\eta) = 1$$

where η is a number between $-\epsilon$ and ϵ; since $G(t)$ as well as $\dot{G}(t)$ are equal to zero for $t < 0$, one has $G(-\epsilon) = \dot{G}(-\epsilon) = 0$. Moreover, since $G(t)$ is continuous at $t = 0$, $G(\epsilon)$ and $G(\eta)$ go to 0 along with ϵ. Thus:

$$\dot{G}(0) = 1/m$$

In other words, if m is kept fixed all Green's functions have the same derivative at $t = 0$ irrespective of the values of b and c.

2.4 Dynamic equations: higher orders

For the Green's function defined in equation (2.4) the derivative experiences a jump at $t = 0$: it is equal to zero for $t < 0$ and jumps to $1/m$ at $t = 0^+$. For some markets (such as real estate markets) such a behavior is at variance with observation, in the sense that at the beginning of the bubble the price begins to rise slowly which rather corresponds to a Green's function with a continuous derivative at $t = 0$. Such behavior can be described by higher-order equations. For example fig. 8.1b corresponds to a fourth-order equation of the form:

$$\left(\frac{d^2}{dt^2} + 2b_1 \frac{d}{dt} + c_1 \right) \left(\frac{d^2}{dt^2} + 2b_2 \frac{d}{dt} + c_2 \right) G(t) = \delta(t)$$

In this case $\dot{G}(0^+) = \ddot{G}(0^+) = 0$ (appendix A). Such an equation also has the advantage that it can describe a peak for which the up-going phase is longer than the down-going phase, as is indeed observed (see chapter 6).

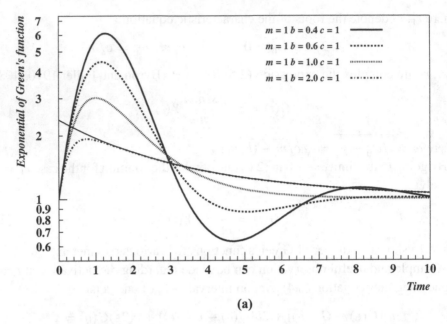

(a)

Fig. 8.1a. Exponential of the Green's function of a second order equation
Notes: The equation reads: $m\ddot{G} + 2b\dot{G} + cG = \delta(t)$; the driving force $\delta(t)$ is an idealization, but the response to a more realistic rectangular pulse would have a similar shape. The graph represents the function: $p(t) = \exp[G(t)]$; with a logarithmic vertical scale the shape of the curve for $p(t)$ is of course the same as the one for $[G(t)]$ with a non-logarithmic scale. The two highest curves correspond to light damping; the third corresponds to critical damping ($b = \sqrt{c}$); the fourth corresponds to heavy damping. The thin line represents the exponential term with the smallest (in absolute value) characteristic root; it shows that in the case of heavy damping the second-order Green's function is not very different from the first-order Green's function. Incidentally it can be noted that the function with critical damping falls off more rapidly than those for heavy damping.

However for the questions discussed subsequently the precise shape of the peaks makes little difference. For the sake of simplicity we will therefore use the second-order equation (2.4).

2.5 *Light or heavy damping?*

Green's functions for light, critical ($b = \sqrt{c}$) and heavy damping are shown on fig. 8.1a. For heavy damping the two roots of the characteristic equation are real and, as far as absolute values are concerned, one is much larger than the other. In short, for heavy damping the Green's function is fairly similar to the case of the first-order equation. This is confirmed on fig. 8.1(a) by the fact that the curve corresponding to the smallest root (fine line) is almost identical to the Green's function except in the vicinity of $t = 0$. Thus, for reasons already mentioned in the previous paragraph the heavy damping case will be of little interest in the present study.

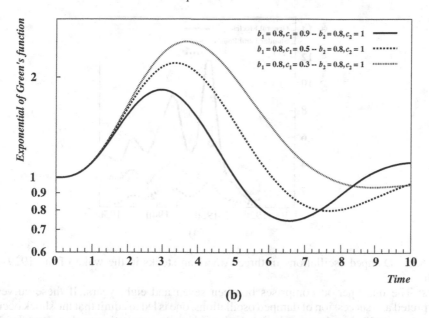

(b)

Fig. 8.1b. Exponential of the Green's function of a fourth-order equation
Notes: In contrast to fig. 8.1a the first and second derivatives are continuous at $t = 0$, a feature which is in line with price peaks for wheat or real estate markets. The curves correspond to the solution of the equation:

$$(d^2/dt^2 + 2b_1 d/dt + c_1)(d^2/dt^2 + 2b_2 d/dt + c_2)G(t) = \delta(t)$$

The calculation is outlined in appendix A.

At this point two comments are in order.

- According to one of the assumptions made above the proportion of investors with respect to users increases along with the price of an item. Therefore one may expect the proportion of investors to increase during the ascending phase of a speculative bubble. In order to describe such an effect one would have to consider either a linear equation whose coefficients change over the course of time or (more appropriately) an equation whose coefficients depend upon the price, that is to say a non-linear equation. In what follows we restrict ourselves to the linear case leaving the exploration of non-linear equations to a subsequent publication.
- In the light damping regime the main price peak is followed by smaller replicas. A natural question is whether such replicas can actually be observed in real price series. When one looks at very long price series in which there are many peaks, such as several century-long wheat price series, one observes in a number of cases that major price peaks are followed by smaller ones. Such was the case in Cologne in the years following 1675, 1699, 1711, 1772; or in Toulouse in the years following 1572, 1630, 1695, 1720, 1812. Naturally since these price movements contain a high level of noise, it cannot be excluded that the replicas were due to exogenous shocks.

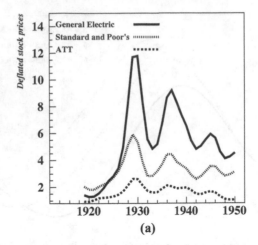

(a)

Fig. 8.2a. Damped oscillations in three American stocks in the wake of the 1929–1930 crash

Notes: The quasi-period comprises between seven and eight years. If these curves are interpreted as a succession of damped oscillations, one is led to admit that the shock occurred in 1927 or 1928; in other words the sections of the curves prior to that date should be considered as an upward trend which does not belong to the speculative peak.

Source: Commercial and Financial Chronicle (various issues).

A somewhat cleaner observation is provided by the behavior of the auto-correlation function. It is well known that in terms of periodicity the auto-correlation function has the same overall behavior as the process itself (Priestley 1981, 131). An analysis of nineteenth-century wheat prices shows that they exhibit damped oscillations, the length of a quasi-period being of the order of 2.5 years (Roehner 1995, 360).

Fig. 8.2(a) shows a number of oscillation replicas in the wake of the price peak of 1929: ATT is characterized by a low up-going trend rate while General Electric is characterized by a large up-going trend rate, the Standard and Poor's Index ranking in between. Figure 8.2(b) shows a model simulation in which these oscillations are interpreted as recurrent (damped) occurrences of the 1929 price peak. Of course it cannot be completely excluded that the fluctuations in figure 8.2a are due to exogenous shocks, but their interpretation as damped oscillations is quite appealing.

3 Implications

We now examine the characteristics of the peaks, beginning with the relationship between their amplitude and the time at which they occur.

3.1 Amplitude versus duration of the ascending phase

Since we are mainly interested in the light damping regime, $\gamma \ll \omega_0$, the oscillations are quasi-periodic and the first maximum occurs approximately when the sinus in

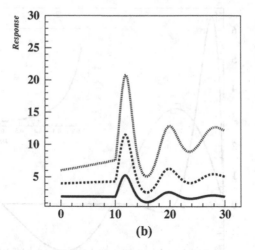

(b)

Fig. 8.2b. Three damped oscillations adjusted qualitatively to the stock price curves
Notes: The parameters of the second-order equation are the same for all three curves: $m = 1$, $b = 0.15$, $c = 0.04$; only the trends are different namely: 0, 0.5 percent and 1.5 percent respectively.

(2.6) is equal to 1, that is to say for:

$$t \simeq \frac{1}{4} \frac{2\pi}{\omega}$$

In chapter 6, this duration was denoted T_{up}; in terms of the parameters of the equation the previous equation reads:

$$T_{\mathrm{up}} \simeq \frac{\pi}{2} \sqrt{\frac{m}{c}} \tag{3.1}$$

Substituting t into the Green's function, one gets the ordinate G_m of the first maximum:

$$G_m \simeq \frac{1}{\sqrt{mc}} \exp\left[-\frac{\pi}{2} \frac{b}{\sqrt{mc}}\right] \tag{3.2}$$

Due to the condition $\gamma/\omega_0 \ll 1$ the exponent will be small and G_m will decrease as m increases.

If we now consider (3.1) and (3.2) as parametric equations defining T_{up} and G_m as functions of m one obtains a relationship of the form $G_m = f(T_{\mathrm{up}})$. In order to make contact with the definitions used in previous chapters we are rather interested in the amplitude A, which is the ratio of the peak price to the initial price. With the definition $p = e^G$ and due to the fact that $G(0) = 0$ one gets $A = e^{G_m}$; this function is represented in fig. 8.3(a) (inset).

It turns out that in order for $A = f(T_{\mathrm{up}})$ to be a decreasing function, as was observed for S-class items, it is the parameter m which has to be changed. In contrast,

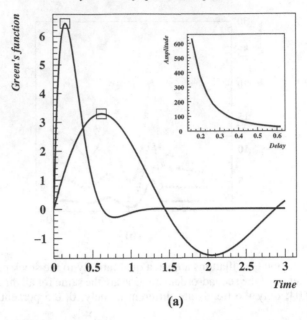

(a)

Fig. 8.3a. Relationship between peak amplitude and delay
Notes: The graph shows the Green's functions corresponding to $m = 0.01, b = 0.05, c = 0.5$ (highest curve) and $m = 0.1, b = 0.05, c = 0.5$ (lowest curve); the inset shows the relationship between peak amplitude (for $p = e^G$) and the time of occurrence of the first maximum, as m varies between 0.001 and 0.1: the smallest values of m correspond to large amplitudes and small delays.

changing either b or c results in a positive correlation between T_{up} and A (fig. 8.3(b)). This implies that there should be a relationship (or at least a positive correlation) between m and the proportion $1 - k$ of users. A qualitative comparison between equation (2.4) and evidence for property prices in Britain is given in fig. 8.3(c).

3.2 Peak amplitude and proportion of investors

In the first section we have introduced the idea that the amplitude of speculative peaks is determined by the proportion of speculators among the buyers. Unfortunately this proportion is seldom known with any precision, and in order to justify that assumption one has to resort to indirect arguments.

For instance one can observe that price peaks are almost non-existent for banana prices, are of moderate size ($A < 2$) for wheat prices, and of an amplitude of 5 (or more) for stock prices. For stock prices there are of course no users which means that $k = 100\%$; in contrast, speculation is almost impossible for bananas because these fruits cannot be stored for more than one or two months (and their storage is fairly costly). As a result nobody will buy bananas with the purpose of holding

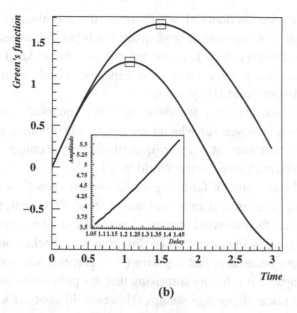

(b)

Fig. 8.3b. Relationship between peak amplitude and delay
Notes: The graph shows the Green's functions corresponding to $m = 0.5, b = 0.01, c = 0.05$ (highest curve) and $m = 0.5, b = 0.01, c = 1$ (lowest curve); the inset shows the relationship between peak amplitude (for $p = e^G$) and the time of occurrence of the first maximum, as c varies between 0.05 and 1.

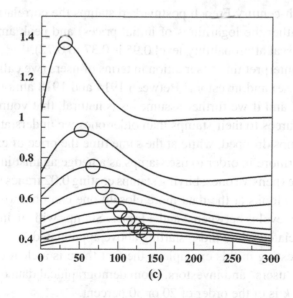

(c)

Fig. 8.3c. Amplitude and time of occurrence of price peaks
Notes: The parameters were selected in order to obtain (at least qualitatively) a set of curves comparable to the one for property prices in the UK (see fig. 7.1b): b is equal to 0.01, c is equal to 0.5 and m varies between 0.04 (highest peak) and 0.5 (flat peaks). These different values of m correspond to different proportions of investors.

them until their price has doubled, which means that k is almost equal to 0. Wheat is an intermediate case; storage of large quantities is possible albeit not easy (grains have to be ventilated regularly). Therefore it is possible to keep large inventories for several months if a price increase is expected, which means that k will be somewhere between 0 and 100 percent.[2]

Unfortunately the previous arguments are more qualitative than quantitative. There is one case, however, for which (due to special circumstances) it is possible to state the previous argument in a more quantitative way: it concerns the speculation in postage stamps in France during World War I.

World War I was a time of fairly high inflation: between 1914 and 1919 French prices on average increased at an annual rate of about 20 percent; more precisely in the first three years the rate was between 15 percent and 20 percent and in 1918 and 1919 it was in the 25–30 percent range. Inflation was checked only in 1921 when the price increase became in fact negative (-12 percent). On account of what we observed in chapter 5 it is hardly surprising that this period was marked by a strong increase in the price of postage stamps. However, in contrast to what happened during World War II, the increase concerned only the most expensive stamps; common stamps in fact experienced a decrease. That situation is summarized in fig. 8.4. Between 1913 and 1919 the (deflated) price of the expensive stamp (13.8 francs in 1913) was multiplied by 3.1, whereas the price of the common stamp (0.02 franc in 1913) was divided by 2.5. The same observation holds for a large sample of stamps: for 57 nineteenth-century French postmarked stamps the correlation between initial prices (or rather the logarithms of initial prices) and peak amplitudes is 0.57 (confidence interval at probability level 0.95 is 0.37 to 0.72).

How can we interpret this observation in terms of users (we call them collectors in the present case) and investors? Between 1914 and 1918 almost all young men were mobilized, and if we further assume, as is natural, that young collectors can devote less resources to their stamps than older ones we understand why the price of common stamps dropped, while at the same time the price of expensive stamps climbed. Furthermore in order to use stamps as a hedge against inflation one must necessarily buy expensive ones; buying stamps costing 0.02 francs would be a waste of money. If one looks at the data more closely one can observe that on average the critical price level is between 2 and 3 francs. Stamps with an initial price below that level depreciated while those starting above appreciated.

What is interesting in this example is the fact there is a clear contrast between the behavior of "users" and investors. From demographical data one may estimate that in this case k is of the order of 20 or 30 percent.

---

[2] It is true that the volume of trade in futures markets is much larger than the volume of trade in spot markets. However, there are good reasons to believe that, except in special circumstances, price trends are rather determined on spot markets.

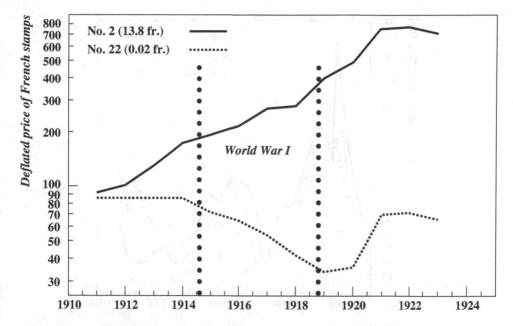

Fig. 8.4. Price of nineteenth-century postage stamps during World War I
Notes: Vertical scale: price expressed in 1910 French francs. The graph shows the price
movements for an expensive stamp (solid line) and for a common stamp (dashed line)
during World War I; while the first increased due to investors' demand, the second fell
because most small collectors were drafted.

3.3 Synchronization effects

The correlation between the prices of different items is a variable of great importance
for consumers. For instance, if the price of corn increases but the price of soya beans
remains stationary, the demand (for instance for cattle feed) will shift from the first
to the second, whereas such a substitution becomes impossible if the two prices are
strongly correlated. It is of importance to observe that the correlation between the
prices p_A and p_B of two items is by no means constant over the course of time: p_A
and p_B may move almost independently for several years and suddenly become
highly correlated when both items experience a price peak.

At first sight, the fact that price peaks tend to synchronize price fluctuations
could seem a trivial and fairly tautological statement, for it is obvious that if both
p_A and p_B experience a peak at the same moment they are *ipso facto* correlated.
But a detailed observation (fig. 8.5(a)) also shows that the synchronization is main-
tained for about three to four years after the price peak, that is a long time after
the exogenous shock that produced the peak has vanished. This is a typical dynamical effect. Fig. 8.5(b) shows that the Green's function of a discrete second-
order equation containing a noise term exhibits the same feature. More precisely

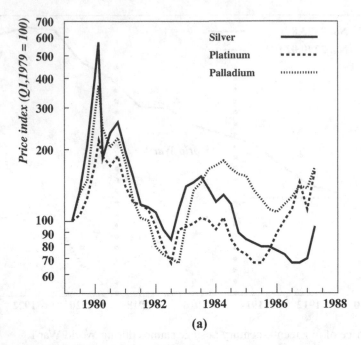

(a)

Fig. 8.5a. Synchronization of price fluctuations by an exogenous shock
Notes: The prices of silver, platinum and palladium peaked simultaneously at the beginning
of 1980; their movements remained correlated for about one quasi-period (four years), but
subsequently the three prices tended to move fairly independently one from another. This
example illustrates a number of similar cases, such as gold–silver which both peaked in
early 1980 but moved almost independently after 1996; or export prices of American and
Thai rice which both peaked in mid 1993 but moved almost independently after 1996.
Sources: The Economist (April 25, 1987), Chalmin (1999).

after the peak the synchronization between the three curves subsists approximately
over one quasi-period. Subsequently the synchronization progressively wanes and
the three prices resume the independent fluctuations they have had prior to the
peak.

Naturally when the dynamical parameters (i.e. the quasi-period) of the different
items are very different the synchronization dies out more rapidly. In the three examples mentioned in the caption of fig. 8.5(a), namely silver–platinum–palladium,
gold–silver, American rice–Thai rice the goods are sufficiently similar to have fairly
close dynamical characteristics.

Figure 8.5(b) was the first instance in this chapter where we considered a
stochastic second-order process; this aspect, which is obviously needed in order to
explain properties pertaining to standard deviations or distribution functions, will
be further developed in the next chapter.

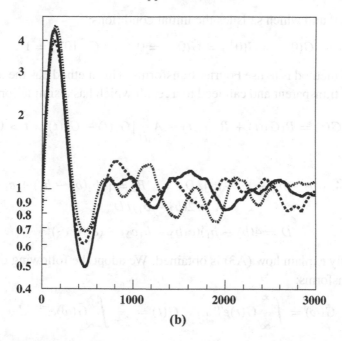

(b)

Fig. 8.5b. Three realizations of the stochastic Green's function of an auto-regressive process
Notes: The equation reads $a_0 X_t + a_1 X_{t-1} + a_2 X_{t-2} = \delta_{t0} + \epsilon_t$ where ϵ_t represents a Gaussian noise term with zero mean and standard deviation equal to 0.5; and $a_0 = 3434.15$, $a_1 = -6847.78$, $a_2 = 3413.79$; these values were obtained by discretization of the continuous case corresponding to the parameters: $m = 0.34$, $b = 0.01$, $c = 0.5$. The horizontal scale is expressed in number of steps.

Appendix A: Green's function for a fourth-order equation

There are two motivations for the present study: (i) the derivative of the Green's function should be continuous at $t = 0$ in order to better fit the shape of real price peaks; (ii) A fourth-order equation is a natural generalization of the second-order equation, in the sense that we consider a class of equations which is a composition of two second-order operators, namely:

$$L_i(d/dt) = \frac{d^2}{dt^2} + 2b_i \frac{d}{dt} + c_i \quad i = 1, 2 \qquad L_1(d/dt)L_2(d/dt)G(t) = \delta(t)$$

$$(A1)$$

As is shown by Routh's theorem (Chiang 1984, p. 546), if the parameters b_i and c_i are positive the stationarity condition is ensured for each second-order process and hence for the global fourth-order process as well.

In order to find the causal solution to equation (A1) there are basically two methods. The first consists in determining the solution of equation (A1) with $\delta(t)$

replaced by 0 and which satisfies the initial conditions:

$$G(0^+) = \dot{G}(0^+) = \ddot{G}(0^+) = 0, \qquad G^{(3)}(0^+) = 1 \qquad (A2)$$

The second method is to use Fourier transforms. This method has the advantage of being more transparent and can lead to a result which has a simple form, namely:

$$G(t) = B_1 G_1(t) + B_2 G_2(t) - A \frac{d}{dt}[G_1(t) - G_2(t)] \quad t > 0 \qquad (A3)$$

where:

$$B_1 = [-4b_1(b_2 - b_1) + c_2 - c_1]/D, \qquad B_2 = [4b_2(b_2 - b_1) + c_1 - c_2]/D,$$
$$A = 2(b_2 - b_1)/D$$

$$D = 4(b_2 - b_1)(c_1 b_2 - b_1 c_2) + (c_1 - c_2)^2$$

Let us briefly explain how (A3) is obtained. We adopt the following definitions of Fourier transforms:

$$\tilde{G}(\omega) = \int_{-\infty}^{\infty} G(t)e^{it\omega}dt \quad G(t) = \frac{1}{2\pi} \int_{-\infty}^{\infty} \tilde{G}(\omega)e^{-it\omega}d\omega \qquad (A4)$$

thus:

$$\mathcal{F}[\dot{G}(t)] = -i\omega\mathcal{F}[G(t)] \qquad (A5)$$

Transforming equation (A1) gives:

$$\tilde{G}(\omega) = \frac{1}{L_1(-i\omega)L_2(-i\omega)}$$

The right-hand side can be decomposed in the standard way:

$$\frac{1}{L_1(-i\omega)L_2(-i\omega)} = \frac{A_1\omega + B_1}{L_1(-i\omega)} + \frac{-A_1\omega + B_2}{L_2(-i\omega)} \qquad (A6)$$

The above expression can be written in the form:

$$\frac{1}{L_1(-i\omega)L_2(-i\omega)} = \frac{B_1}{L_1(-i\omega)} + \frac{B_2}{L_2(-i\omega)} - A(-i\omega)\left(\frac{1}{L_1(-i\omega)} - \frac{1}{L_2(-i\omega)}\right) \qquad (A7)$$

where $A = A_1/i$

At this point we note that $1/L_i(-i\omega)$, $i = 1, 2$ represents the Fourier transform of the Green's function $G_i(t)$ of the operator $L_i(d/dt)$. Furthermore it results from (A5) that $-i\omega/L_i(-i\omega)$ is the Fourier transform of the derivative dG_i/dt. This leads immediately to the solution (A3).

Two remarks are in order: (i) The above calculation can easily be generalized to an arbitrary number of second-order operators. (ii) It is of interest to note that for

$b_1 = b_2$ the solution takes the particularly simple form:

$$G(t) = \frac{G_1(t) - G_2(t)}{c_2 - c_1} \tag{A8}$$

Finally let us check that the solution (A3) indeed satisfies the initial conditions (A2). Differentiating (A3) leads to:

$$G^{(k)}(0) = B_1 G_1^{(k)}(0) + B_2 G_2^{(k)}(0) - A\left[G_1^{(k+1)}(0) - G_2^{(k+1)}(0)\right] \quad k = 1, 2, 3 \tag{A9}$$

In order to determine the $G_i^{(k)}(0)$ let us, for instance, assume that the (b_i, c_i) correspond to the quasi-periodic regime; in this case:

$$G_i(0^+) = 0, \qquad \dot{G}_i(0^+) = 1, \qquad \ddot{G}_i(0^+) = -2b_i,$$
$$G_i^{(3)}(0^+) = 4b_i^2 - c_i, \qquad G_i^{(4)}(0^+) = 4b_i(c_i - 2b_i^2)$$

Substituting into (A9) leads to the initial conditions (A2).

9

Theoretical framework: implications

In the present chapter we explore a number of implications and applications of the theoretical framework delineated in the previous chapter. Several of these applications require a stochastic framework; nevertheless we will avoid technicalities as far as possible in order to keep this chapter in line with the rest of the book. As far as the more technical sections are concerned readers may, if they wish, jump directly to the conclusions without losing the gist of the argument.

In the first section we consider an application which, although very simple mathematically, has interesting consequences, which can be summarized a follows. During a speculative episode (for instance the bull market of 1920–1929) the higher the price of a stock climbed during the rising phase the better it resisted during the subsequent bear market. This regularity, which will be referred to as the resilience pattern, is not restricted to stocks but can also be observed for other items, such as stamps, houses, or apartments.

The second section is devoted to what the Boston group (Plerou, Amaral, Gopikrishnan, Meyer, and Stanley 2001) called the breakdown of scaling. If one considers the distribution of the logarithms of stock price changes $\Delta p(t, \theta) = \ln p(t) - \ln p(t - \theta)$ one observes that the width of the distribution remains constant (provided the differences are normalized by dividing them by the standard deviation σ_p) for θ comprised between a few minutes and one day; however, for θ comprised between two weeks and several months the width of the distribution decreases.

In the third section we analyze the ensemble coefficient of variation during a speculative episode; this gives us the opportunity to emphasize that, for non-stationary processes, time averages have only little significance and it becomes essential to consider the statistical properties of ensembles of items, e.g. samples of stocks, sets of houses, or collections of stamps. In recent times this avenue has been extensively explored by Rosario Mantegna and his colleagues (Lillo and Mantegna 2000, 2001); in particular, they analyzed the correlation range of second moments and showed that the order of magnitude is the same for the NYSE and NASDAQ markets.

198

In a sense the models introduced in the previous chapter (we call them GC models which stands for Green-comparative) can be considered as two-body problems in the extended sense we have given to that expression in the first chapter. Indeed they describe the interaction between a dynamical system and an exogenous impulse. However, such a framework becomes insufficient when one tries to describe the interaction between different items or different markets. For that reason in the fourth section we consider a generalization to spatially interacting markets which describes U-class markets; this framework will be called the stochastic spatial arbitrage (or more briefly SSA) model.

There is another potential application which needs to be mentioned although we will not be able to devote much space to it in the present chapter. It concerns the interpretation of the so-called log-periodic pattern of speculative peaks. This avenue was pioneered and systematically investigated by Didier Sornette and his collaborators (Sornette, Johansen, and Bouchaud 1999, Johansen and Sornette 1999a,b). It provides a short-term description of the shape of speculative peaks which, in a medium-term perspective, follow the sharp peak–flat trough pattern. The main prediction is the appearance of oscillations which become more and more rapid as the price approaches the turning point.

Such a pattern has a natural interpretation in the GC framework of the previous chapter. By using a fourth-order equation, which mixes slow and fast frequencies, one can describe a peak, which is "decorated" with short-term fluctuations. Moreover the GC framework predicts that, as the price level increases, the proportion of investors grows, which in turn results in a shortening of the response time of the system. In other words, as the market converges toward the turning point, the GC framework predicts more and more rapid fluctuations, which is precisely what is observed in the log-periodic pattern.

One difference is that the log-periodic pattern predicts an infinite frequency in the close vicinity of the turning point. Whereas in the GC framework there is an upper bound to the frequency, which is defined by the response time of a population of traders that would comprise 100 percent investors. On stock markets this upper bound is fairly high but on real estate markets it is much lower. The application of the GC framework to log-periodic oscillations will be explored in greater detail in a subsequent publication.

1 The resilience effect

1.1 Description

We consider a speculative episode for S-class items. To make the discussion more concrete one may consider the example of the 1929 stock price bubble in the United

States. We denote by $p_1(k)$ the price level of stock number k (e.g. General Electric) at the start of the peak t_1, by $p_2(k)$ the price of the same stock at peak time t_2 (e.g. $t_2 = 1929$), and by $p_3(k)$ its price at the end of the falling phase (e.g. $t_3 = 1932$). We further introduce the:

$$\text{peak amplitude: } A(k) = p_2(k)/p_1(k) \quad \text{and}$$
$$\text{bottom amplitude: } B(k) = p_3(k)/p_1(k)$$

$A(k)$ and $B(k)$ can be defined for any stock provided it was already in existence at time t_1 and is still in existence at time t_3. Below we will see that there is a definite relationship between $A(k)$ and $B(k)$ and with a correlation of the order of 0.75 the following relationship holds:

$$B = aA + b + \epsilon \quad \epsilon : \text{random noise} \tag{1.1}$$

where A and B denote the random variables of which $A(k)$, $B(k)$ are realizations; the coefficient a is usually of the order of 0.5. The fact that a is positive means that the higher the peak amplitude, the larger the bottom amplitude. In other words, the higher the price of a stock climbs during the rising phase (bull market) the better it resists during the subsequent falling phase (bear market); for this reason the regularity expressed by equation (1.1) will be referred to as the resilience pattern.

Before giving more detailed statistical evidence we first examine how this effect can be interpreted in the GC framework of the previous chapter.

1.2 Interpretation

The idea behind the interpretation is summarized in fig. 9.1(a),(b). The first figure displays annual price data for four American companies over a 45 year interval. The bull market lasted from the early 1960s to the late 1960s, whereas the bear market lasted from the late 1960s to the mid 1970s. One also observes that in spite of many irregular price fluctuations each stock has a definite price trend. That idea has been implemented in the model and led to fig. 9.1(b). As in chapter 8 the price is defined by $p = e^G$ where G is the solution of equation (2.4) in chapter 8, and in addition a trend term at has been added on the right-hand side. As a result the price p has a constant increase rate ρ. If we denote by $A_0 = (p_2/p_1)_{\rho=0}$ and $B_0 = (p_3/p_1)_{\rho=0}$ the ratios p_2/p_1 and p_3/p_1 when there is no trend (i.e. $\rho = 0$) one has:

$$A(k) = A_0 + \exp[\rho(k)T_{\text{up}}], \quad B(k) = B_0 + \exp[\rho(k)(T_{\text{up}} + T_{\text{down}})] \tag{1.2}$$

As a result the regression coefficient a will be given by:

$$a = \frac{\Delta B}{\Delta A} = \exp[\rho(k)T_{\text{down}}] > 0 \tag{1.3}$$

We find that a is indeed positive as implied by the resilience effect. From the

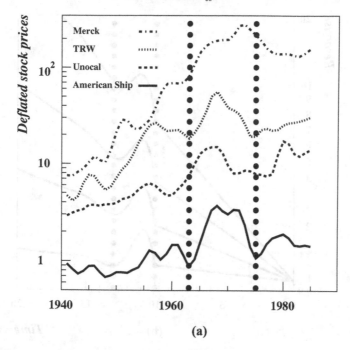

Fig. 9.1a. Share prices for four American companies
Notes: Merck is a global research-driven pharmaceutical company. TRW provides advanced technology products for the automotive, aerospace and information system industries. Unocal is an oil and gas production company. These companies have been selected in order to display a wide range of growth rates: the price of American Ship stock has negligible growth rate while for Merck it is about 10 percent. The figure shows that for high-growth companies the ratio p_3/p_1 is larger than for slow-growth companies. As we are only interested in the price changes in the course of time, the curves for Merck and TRW were shifted upward for the sake of clarity.
Sources: Common Stock Price Histories (1987).

simulations in fig. 9.1(b) it is clearly visible that the effect of a bear market is softened for high-growth items. In other words the simple trend model is qualitatively in agreement with the evidence.

However at a more quantitative level one can observe that the coefficient a predicted by (1.3) is always larger than 1, whereas the evidence rather suggests a coefficient which is smaller than 1. This means that ΔA should in fact be larger than the value that results from (1.2); this would be the case if the amplitude A_0 increased with the trend instead of being the same for all curves. This idea is in fact quite consistent with our assumptions in chapter 8; indeed for a high-growth company (as for expensive stamps or houses) there will be more investors than for a slow-growth company and hence a greater price increase is expected.[1]

---

[1] This version of the price multiplier effect is indeed confirmed by the evidence, see in this respect Roehner 2001a, 148.

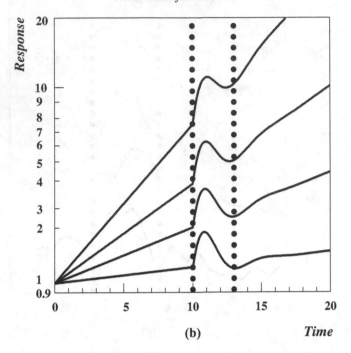

Fig. 9.1b. Simulation of price movements in the GC framework of chapter 8
Notes: The curves correspond to $p = e^G$ and incorporate a growth trend. The parameters of the second-order equation are the same for the four curves, namely $b = 0.7$, $c = 2.2$; the curves only differ by their trends: 4, 14, 25, and 40 percent respectively. The important observation is that higher price increase trends lead to higher p_3/p_1 ratios.

We now present the statistical evidence which supports the resilience pattern.

1.3 Statistical evidence

First we briefly explain the methodology. One of the main problems is the identification of t_1 and t_3 (the determination of the peak time t_2 is of course easy). A peak will be delimited in two steps; at the level of the whole market the price peak will be identified by using an annual index, such as the S&P 500. Then, at the level of each individual company, t_1 will be selected as the first year for which the annual price change is positive, and t_3 as the last year for which the annual price change is negative. That procedure leads to the results summarized in table 9.1.

It must be emphasized that the resilience pattern rules the medium-term price behavior in the vicinity of a price peak but it does not give any information about the long-term behavior. A case in point is provided by the American company Columbia Gas Systems. The stock price of that utility increased tremendously, from 100 to 1,000, during the 1921–1929 bull market; between 1930 and 1932 it dropped to 200 and continued to fluctuate around that price between 1932 and 1936. But in the late 1930s it dropped to 40 and subsequently never really recovered. In other

Table 9.1. *The resilience pattern: relationship between peak amplitude* (A) *and bottom amplitude* (B): $B = aA + b$

	Market	Peak	Number of items	a	b	Correlation
Stocks						
1	NYSE	1929 Oct.	85	0.40 ± 0.05	-0.27 ± 0.24	0.87
2	Paris	1929 Feb.	19	0.44 ± 0.17	-0.03 ± 0.29	0.76
3a	Tokyo	1989 Oct.	26	0.39 ± 0.11	0.27 ± 0.10	0.80
3b	Tokyo	1989 Oct.	26	0.40 ± 0.09	0.35 ± 0.13	0.87
Stamps						
4	France	1944	56	0.10 ± 0.08	0.77 ± 0.10	0.33
Real estate						
5	Paris	1990	20	0.43 ± 0.16	0.73 ± 0.04	0.79
6	Paris	1990	5	0.14 ± 0.26	1.1 ± 0.1	0.52
7	Britain	1989	11	-0.10 ± 0.17	1.4 ± 0.05	-0.38
	Average	(except 7)		0.32		0.68

Notes: All prices used in the regression are real (i.e. deflated) prices. Case (3a) refers to $t_1 = 1985$, while case (3b) refers to $t_1 = 1980$; the comparison shows that a does not depend upon t_1 in a critical way. For some reason yet to be understood the housing bubble in Britain does not follow the resilience pattern (negative correlation).

Sources: 1: *Common Stock Price Histories* (1987); 2: *Annuaire Statistique de la France* (*Rétrospectif* 1966), 541; 3a,b: *Japan Statistical Yearbook* (various years); 4: Massacrier (1978); 5,6: Conseil par des Notaires (Dec. 23, 1991) and Chambre des Notaires; 7: Halifax index.

words we have here an example of a fast-growth company which experienced one huge peak but lost most of its value in the long run.

2 Breakdown of scaling

Since the pioneering studies by Mandelbrot (1963) and Fama (1965), the determination of the asymptotic distribution of stock price changes has been a much disputed issue. The econophysics community and in particular the Boston group devoted much attention to this question (Gopikrishnan *et al.* 1998, Plerou *et al.* 2001). The main issue is whether or not the distribution belongs to the class of stable Levy laws (for more detail see Mantegna and Stanley 2000). Here, we will consider another, albeit related, question.

The Boston group has shown (Plerou *et al.* 2001: figures 6(b) and 9(a)(b)) that the distribution of price changes normalized by dividing them by the standard deviation is unchanged for time intervals comprised between 5 and 320 minutes. But for increasing time intervals, ranging from several days to several years, the width of the distribution decreases or, to say it differently, it falls off more rapidly (Fig. 9.2). In the

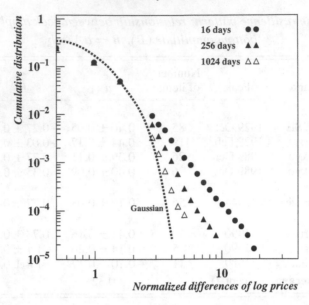

Fig. 9.2. Cumulative distribution of log-price changes on the NYSE
Notes: The graph represents the positive tail of the cumulative distribution of $\ln[p(t)] - \ln[p(t - \theta)]$ for different values of the time lag θ. The data are share prices for individual companies and cover the period 1962–1996. As θ increases the tails converge toward the Gaussian distribution, which means that stock prices almost follow a log-normal distribution.
Source: Adapted from Plerou *et al.* (2001).

following paragraphs we examine this question in the framework of a second-order auto-regressive process, which is the analog in discrete time of the second-order differential equation used in chapter 8. Our analysis will highlight a relationship between the response time of the system and the width of the distribution. Intuitively such a relationship could of course be expected, but here we will get a quantitative statement. To begin with, for the sake of simplicity, we examine the first-order case.[2]

2.1 First-order process

The main steps of the calculation are the same in the first- and second-order cases, but technically it is simpler to explain them for a first-order equation; thus we start with a process of the form:

$$X_t = aX_{t-1} + \epsilon_t \tag{2.1}$$

where ϵ_t is a random noise of zero mean and standard deviation σ_ϵ.

---

[2] I would like to express my gratitude to my colleague Benoît Douçot for fruitful discussions on the issues considered in this section.

Equation (2.1) is the discrete version of a first-order stochastic differential equation:

$$\frac{dX}{dt} = aX(t) + \epsilon(t) \tag{2.2}$$

Equation (2.1) has two main advantages over equation (2.2). (i) The definition of the continuous random noise $\epsilon(t)$ is a tricky matter from a mathematical point of view (see for instance in this respect Grimmet and Stirzaker 1982) whereas the definition of ϵ_t does not involve the same difficulties. (ii) Computer simulations necessarily use equation (2.1); in addition a discrete framework parallels the time series that are given in statistical sources.

We assume that the process (2.1) begins at $t = 0$, which implies that $X(t < 0) = 0$, $\epsilon(t < 0) = 0$; for $t = 0, 1, \ldots$ equation (2.1) reads:

$$X_0 = \epsilon_0, \ X_1 = aX_0 + \epsilon_1, \ldots$$

which gives the standard result:

$$X_t = \epsilon_t + a\epsilon_{t-1} + a^2\epsilon_{t-2} + \cdots + a^{t-1}\epsilon_1 + a^t\epsilon_0$$

We are interested in the distribution of successive changes, that is to say:

$$g_t \equiv X_t - X_{t-1} = \epsilon_t + (a-1)\epsilon_{t-1} + a(a-1)\epsilon_{t-2} + \cdots + a^{t-1}(a-1)\epsilon_0$$

If the ϵ_i are independent Gaussian variables, g_t will also be a Gaussian variable which will be completely defined by its mean and standard deviation σ_g:

$$\sigma_g^2 = \left[1 + (1-a)^2 \left(1 + a^2 + \cdots + (a^2)^{t-1}\right)\right]\sigma_\epsilon$$

$$\sigma_g^2 = \left[1 + (1-a)^2 \frac{1 - (a^2)^t}{1 - a^2}\right]\sigma_\epsilon^2 \tag{2.3}$$

It is known that the stationarity condition for equation (2.1) is $|a| < 1$; under this assumption, for t sufficiently large (2.3) becomes:

$$\sigma_g^2 = \left[1 + \frac{(1-a)^2}{1-a^2}\right]\sigma_\epsilon^2 = \frac{2\sigma_\epsilon^2}{a+1} \qquad |a| < 1 \tag{2.4}$$

One observation is in order. If a is close to but smaller than 1, σ_g^2 does *not* become arbitrarily large. This, however, is an exceptional case; as a rule (as will also be seen below) σ_g diverges when the parameters of the process tend toward the limits of the stability domain. The present anomaly is due to the fact that for $a = 1$, X_t is a Brownian motion and, as one knows, the derivative of a Brownian motion has a finite standard deviation.

Let us express (2.4) in an alternative form by introducing the response time τ of the process (2.1); τ can be defined by considering the response of the system to an impulse $\delta_{0,t}$, which is described by its Green's function $G_t = a^t, t \geq 0$, which leads to $\tau = 1/\ln(1/a)$ (for the sake of simplicity we assume that a is positive) which implies that $a = e^{-1/\tau}$. Substituting into (2.4) leads to:

$$\frac{\sigma_g^2}{\sigma_\epsilon^2} = \frac{2}{e^{-1/\tau} + 1} \tag{2.5}$$

For $\tau \to 0$, $\sigma_g^2/\sigma_\epsilon^2 \to 2$, while for $\tau \to \infty$, $\sigma_g^2/\sigma_\epsilon^2 \to 1$; in other words, when the response time of the system increases from 0 to ∞, the width of the distribution of successive changes decreases.

In the breakdown of scaling that was observed by the Boston group they did not consider σ_g/σ_ϵ, for σ_ϵ cannot be estimated, but rather σ_g/σ_X; nevertheless the same conclusion holds. Indeed, as one knows (Priestley 1981, 118):

$$\sigma_X^2 = \frac{1 - a^{2t}}{1 - a^2}\sigma_\epsilon^2$$

which means that (2.3) can be written in the form:

$$\sigma_g^2 = \sigma_\epsilon^2 + (1 - a^2)\sigma_X^2$$

that is to say:

$$\frac{\sigma_g^2}{\sigma_X^2} = 1 + \left(\frac{\sigma_\epsilon}{\sigma_X}\right)^2 - a^2 = 1 + \left(\frac{\sigma_\epsilon}{\sigma_X}\right)^2 - e^{-2/\tau}$$

When τ increases from 0 to ∞, σ_g^2/σ_X^2 decreases from $1 + (\sigma_\epsilon/\sigma_X)^2$ to $(\sigma_\epsilon/\sigma_X)^2$.

How can the breakdown be interpreted from a microeconomic perspective? Time intervals between 5 and 320 minutes correspond to transactions taking place on the same trading day; thus, the scaling property means that the economic agents who respond to exogenous shocks within a trading day have fairly homogeneous behavior. However, the breakdown of scaling for time intervals of the order of several weeks means that the agents who respond to market changes with long delays have a different trading behavior. As one knows from chapter 5, such agents mainly consist in mutual fund shareholders; that these people do not react in the same way as market makers or day traders is hardly surprising.

Remark. Usually when studying the distribution of price changes one considers the difference of the logarithms of prices rather than the price differences; therefore X_t should be interpreted as $\ln p_t$.

2.2 Second-order process

We now consider a process which is the discrete analog of the equation considered in the previous chapter:

$$X_t + a_1 X_{t-1} + a_2 X_{t-2} = \epsilon_t \qquad (2.6)$$

As before we are interested in the distribution of $g_t = X_t - X_{t-1}$. As the calculation is similar to the previous one, we only give the main results. The expression for σ_g/σ_ϵ is:

$$(\sigma_g/\sigma_\epsilon)^2 = \frac{2}{(1 - r_1 r_2)(1 + r_1)(1 + r_2)} \qquad (2.7)$$

where r_1, r_2 are the roots of the equation:

$$r^2 + a_1 r + a_2 = 0$$

In the case of complex roots (2.7) takes the more suggestive form:

$$(\sigma_g/\sigma_\epsilon)^2 = \frac{2}{(1 - r^2)(1 + 2r \cos \omega + r^2)} \qquad (2.8)$$

where $r_{1,2} = r e^{\pm i\omega}$.

From (2.8) it can be seen that σ_g/σ_ϵ diverges as r approaches 1 from below which is consistent with the stationarity condition for equation (2.6). There are here two time constants: one is the quasi-period $\tau_1 = 2\pi/\omega$ and the other is the inverse of the damping rate; for the sake of simplicity (and because the damping rate is of marginal importance as seen in chapter 5) we consider only the effect of a change in τ_1. Equation (2.8) implies that if τ_1 increases, ω decreases, $\cos \omega$ increases, and σ_g/σ_ϵ decreases which parallels the result obtained in the previous paragraph.

Instead of σ_g/σ_ϵ we now consider: σ_g/σ_X. Taking into account that (Priestley 1981, p.128):

$$\sigma_X^2 = \frac{(1 + a_2)\sigma_\epsilon^2}{(1 - a_2)(1 - a_1 + a_2)(1 + a_1 + a_2)}$$

equation (2.7) can be expressed in the form:

$$\sigma_g^2 = \frac{2\sigma_\epsilon^2}{(1 - a_2)(1 - a_1 + a_2)} \implies (\sigma_g/\sigma_X)^2 = 2\left(1 - \frac{-a_1}{1 + a_2}\right) \qquad (2.9)$$

a_2 is the parameter which determines the damping rate and will therefore be supposed to remain constant. The relationship between τ_1 and a_1 is:

$$\cos(2\pi/\tau_1) = \frac{-a_1}{2\sqrt{a_2}} \qquad (2.10)$$

For simplicity we assume that $-a_1 = a_1'$ is positive; now if τ_1 increases, the left-hand side of equation (2.10) increases too, which implies that a_1' increases and through (2.9) σ_g/σ_X decreases, which parallels the result obtained in the first-order case.

One can conclude this discussion by the following statement:

Proposition. The breakdown of scaling of the distribution of price changes observed for intervals of times larger than several days can be interpreted as being the result of an increase in the response time of the system. This interpretation is coherent with the fact that mutual fund shareholders have a longer response time and a different trading behavior than day traders.

3 Ensemble coefficient of variation

In a general way the calculations performed in the framework of probability theory concern ensemble averages. The calculation performed in the previous section provides an illustration. However, statistical estimates usually rely on time averages, but, for non-stationary processes, such as the impulse responses considered in the previous chapter, such a procedure can no longer be used. For instance it would make little sense to compute time averages (either the mean or standard deviation or higher order moments) for the silver–platinum–palladium prices shown in fig. 8.5a, for the results would crucially depend upon the width of the time window. In short, for non-stationary processes the most natural approach is to consider ensemble averages. For instance, it makes sense to compute at any time the average of the silver, platinum, and palladium prices.

As a matter of fact this point is of crucial importance not only in economics but in all the social sciences. In sociology for instance time averages are called longitudinal averages and ensemble averages are called transversal (or cross-national) averages. Thus, the previous argument shows that if one wants to study highly non-stationary events, such as revolutions, the appropriate methodology is to perform ensemble averages over a set of instances.

This kind of consideration led us in chapter 7 to consider the behavior of the ensemble coefficient of variation. In the present section we analyze that behavior analytically. The calculation can be performed in continuous or discrete time; we selected the second procedure. The analysis that follows applies to any auto-regressive process.

If for the purpose of illustration one considers a first-order process, the starting equation reads:

$$X_t - aX_{t-1} = \epsilon_t + h\delta_{t,0} \tag{3.1}$$

In (3.1) ϵ_t denotes a Gaussian noise with zero mean and standard deviation σ_ϵ; the ϵ_t are supposed independent; $\delta_{t,0}$ is the discrete analog of the delta distribution: $\delta_{t,0} = 0$ except for $\delta_{0,0} = 1$.

In the previous section we assumed that $\epsilon_{t<0} = 0$; if we make the same assumption here we will mix two different problems, namely the transient behavior due to the fact that the auto-regressive process begins at $t = 0$ and the transient behavior due to $\delta_{t,0}$. This is of course not what we want; in practice when the shock $\delta_{t,0}$ occurs the system has already been in existence for a long time; which means that one should rather assume that $\epsilon_{t<0} \neq 0$ and disregard the auto-regressive transient regime.

Introducing the Green's function of the auto-regressive process, X_t can be written as:

$$X_t = \sum_{k=-\infty}^{\infty} f_k G_{t-k} \qquad G_{k<0} = 0 \tag{3.2}$$

where $f_k = \epsilon_k + h\delta_{k,0}$

As in the previous chapter the price is defined by:

$$p_t = e^{X_t}$$

Our objective is to obtain $\sigma_p^2 = E(p_t^2) - E^2(p_t)$. If ϵ_t is Gaussian with zero mean and standard deviation σ_ϵ, X_t will be Gaussian too with mean m_X and standard deviation σ_X and p_t will be log-normal with mean m_p and standard deviation σ_p. In terms of the parameters of the Gaussian the mean and standard deviation of a log-normal random variable are given by (Parzen 1960, 348):

$$m_p = \exp\left[m_X + \frac{1}{2}\sigma_X^2\right], \qquad \sigma_p^2 = e^{2m_X + \sigma_X^2}\left(e^{\sigma_X^2} - 1\right) \tag{3.3}$$

Remark. As m_X and σ_X^2 do not in general have the same dimension, it may seem at first sight that the previous expressions violate standard homogeneity requirements. However, for the definition of a log-normal variable to be legitimate, X must be dimensionless.

First we compute $m_X = E(X_t)$; without the term $\delta_{t,0}$, m_X would of course be equal to zero, but the impulse leads to a transient non-zero value $m_X = hG_t$.

In order to get σ_X^2 we compute: $E(X_t X_{t+l})$:

$$E(X_t X_{t+l}) = \sum_{i,j} G_{t-i} G_{t+l-j} E(f_i f_j) \tag{3.4}$$

furthermore:

$$E(f_i f_j) = E(\epsilon_i \epsilon_j) + h[\delta_{j,0} E(\epsilon_i) + \delta_{i,0} E(\epsilon_j)] + h^2 \delta_{i,0} \delta_{j,0}$$

From our assumptions about the noise results that:

$$E(\epsilon_i) = E(\epsilon_j) = 0, \qquad E(\epsilon_i \epsilon_j) = \sigma_\epsilon^2 \delta_{i,j}$$

Thus:

$$E(X_t X_{t+l}) = \sigma_\epsilon^2 \sum_{k=-\infty}^{\infty} G_k G_{k+l} + h^2 G_t G_{t+l} \qquad (3.5)$$

For $l = 0$ equation (3.5) gives:

$$E\left(X_t^2\right) = \sigma_\epsilon^2 \sum_{k=-\infty}^{\infty} G_k^2 + h^2 G_t^2 \qquad (3.6)$$

and:

$$\sigma_X^2 = E\left(X_t^2\right) - m_X^2 = \sigma_\epsilon^2 \left(\sum_{k=-\infty}^{\infty} G_k^2 \right)$$

Substituting into (3.3) one gets for the coefficient of variation $c_p \equiv \sigma_X / m_X$:

$$c_p = \sqrt{e^{\sigma_X^2} - 1} = \left[\exp\left(\sigma_\epsilon^2 \left(\sum_{k=-\infty}^{\infty} G_k^2 \right) \right) - 1 \right]^{1/2} \qquad (3.7)$$

It is of interest to observe that σ_X and c_p are independent of t and h, which means that they are the same whether or not there is an exogenous pulse; however m_p depends on h.

In the case of the first-order process corresponding to equation (3.1), $G_k = a^k$ and c_p has the following explicit expression:

$$c_p = \left[\exp\left(\frac{\sigma_\epsilon^2}{1 - a^2} \right) - 1 \right]^{1/2}$$

In conclusion under the assumptions we have made, the ensemble coefficient of variation does not exhibit any trough or peak (expect of course those due to random fluctuations) in response to an exogenous impulse. This result is indeed confirmed by computer simulations.

How then can one interpret the trough observed for U-class goods and the peak observed for S-class items? The previous model does not take into account the interactions between different markets. In the previous sections we have considered the price movements on separate markets as different realizations of the same stochastic process; as illustrated in fig. 8.5b this means that the price in market A moves independently of the price on market B at least in normal times. For instance in the case of wheat prices in Bavaria the price in Munich is supposed, at least in normal times, to move independently of the price in Lindau (150 km to the south-west

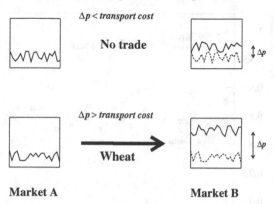

Fig. 9.3. The spatial arbitrage mechanism
Notes: The figure schematically describes the case of two spatially separated wheat markets. The solid lines correspond to price fluctuations on each market; for market B the prices at market A are shown as dashed lines for the sake of comparison. There is no trade between the two markets so long as the price differential remains lower than the transport cost c. Once the price gap becomes larger than c there is an incentive for traders to ship wheat from market A (low price) to market B (high price).

of Munich). This can only be a first approximation however. One can explicitly introduce an interaction between p_A and p_B by replacing the one-dimensional auto-regressive process by a two-dimensional process in the (x, t) plane. We will not pursue that avenue here, but in the next section we present a model which accounts for spatial interactions between markets for U-class goods.

4 The stochastic spatial arbitrage model for U-class goods

The stochastic spatial arbitrage (SSA) model was developed in several previous publications: Roehner 1995 (chapters 3, 4, and 9), 1996, 1999b, 2000b.[3]

The problem can be briefly described as follows. Consider a number of spatially separated wheat markets, how do they interact? An answer is provided by the spatial arbitrage mechanism: two markets A and B do not interact if the price differential between them is smaller than the cost of transport, for in this case traders have no interest in shipping wheat from A to B or from B to A. Once the price gap becomes larger than the transport cost there is an incentive to ship wheat from the market where the price is low to the market where it is high (fig. 9.3). This mechanism establishes an interaction between any set of markets. It is this interaction which is described in the SSA model.

[3] In the first publications it was called the Enke–Samuelson model as a tribute to S. Enke and P. Samuelson who considered the problem of spatially separated markets in the early 1950s. As this designation was found to be rather opaque by many colleagues it was replaced in the 2000b paper by the expression "stochastic arbitrage model."

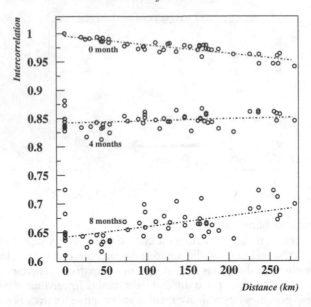

Fig. 9.4. Inter-correlation with a time lag between spatially separated wheat markets
Notes: This figure illustrates one of the most striking predictions of the SSA model. The sample consists of 11 wheat price series (sampling time is one fortnight) for markets located in the center of France during the period 1841–1858. Expressed in 10^{-4} km^{-1} the three slopes are from top to bottom: -1.6, 0.32, and 1.8. The chart only shows the section of the curves for short distances: for sufficiently large inter-market distances the inter-correlations must vanish, but that decrease is not easy to observe because the experimental error bars become fairly wide for large distances.
Source: Adapted from Roehner (1999b).

The SSA model permits numerous predictions about the space-time pattern of prices (spatial price differentials and correlations, correlation length, standard deviation) and about the trade within a set of markets; many of them can be confronted with statistical evidence. One of the most striking predictions is the fact that the price correlation with a time-lag of several months *increases* with inter-market distance (Roehner 1999b). This prediction is unexpected because one is used to the fact that "near things are more related than distant things," a statement which is sometimes called the first law of geography. And yet, this prediction is indeed confirmed by the data (fig. 9.4).

In the present section we want to show that the SSA model is consistent with the pattern between initial prices and peak amplitudes described in chapter 7 for U-class goods. In a continuous variable framework the SSA model takes the form of a second-order partial differential equation driven by a noise term and an impulse (Roehner 1995, chapter 9). For the sake of simplicity we consider here the simple case of a diffusion equation and we leave aside the noise term which is not essential for the present purpose. In other words we consider the Green's function G which

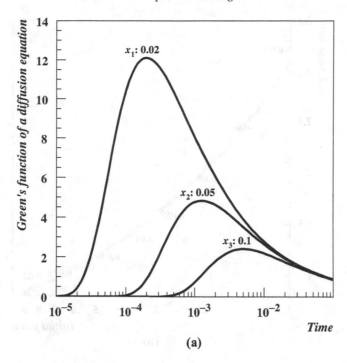

(a)

Fig. 9.5a. Green's functions for the diffusion equation
Notes: At time $t = 0$ there is an impulse at $x = 0$. The graph shows what three observers
located respectively at $x = 0.02, 0.05, 0.1$ would observe: first they see a G which is almost
equal to 0, then the impulse arrives, and finally as the impulse moves further G returns to 0.

satisfies the equation:

$$\partial_t G - a^2 \partial_{x^2}^2 G = \delta(t)\delta(x) \tag{4.1}$$

As one knows the solution is of the form (Tychonov and Samarski 1964):

$$G(x, t) = \frac{1}{\sqrt{\pi a^2 t}} \exp\left(\frac{-x^2}{4a^2 t}\right) Y(t)$$

Three functions $G(x_k, t)$ $(k = 1, 2, 3)$ are shown in fig. 9.5a; they represent what
would be seen by three observers located at x_1, x_2, x_3 respectively. They first see G
increase, reach a maximum, and then decrease back to zero. Given an initial time
t_1 one can define:

$$p_1(x_k) = \exp[G(x_k, t_1)] \quad \text{and} \quad p_2(x_k) = \text{Max}\left[e^{G(x_k, t)}\right]$$

The resulting points are reasonably well fitted (the correlation is 0.92) by a function
of the form:

$$p_2/p_1 = a \ln p_1 + b \qquad a = -1.36, \quad b = 10$$

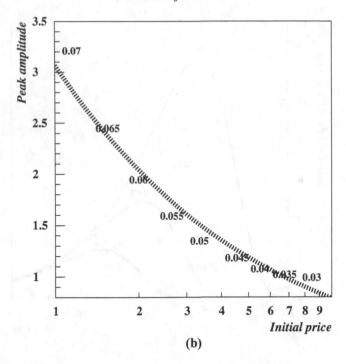

(b)

Fig. 9.5b. Relationship between initial value and amplitude of the peak
Notes: This figure parallels fig. 7.2a. The vertical scale shows the values of Max $G(x_k, t)/$
$G(x_k, t_1)$; the horizontal scale shows the values of $G(x_k, t_1)$. The labels on the curve give the
x-values of the observation points. The dashed line corresponds to the fit function $y = x^{-0.59}$
which can fairly well be approximated by (the correlation is -0.96): $y = -1.04 \ln x + 2.9$.
If one replaces G by e^G the graphic remains very similar but the slope is changed to -1.36
and correlation to -0.92.

   It can be noted that a similar relationship holds whether one considers G (fig. 9.5b)
or e^G.

   The value of a depends upon t_1, but its order of magnitude is consistent with the
estimates given in table 7.1.

   It should be noted that the right-hand side of equation (4.1) can be considered
only as a first approximation, for an exogenous shock usually affects simultaneously a whole region; in other words a right-hand side of the form $\delta(t) \exp(-x^2/L^2)$
(where L is the extent of the region) would be more appropriate; although of
course less tractable analytically. We postpone further development of that idea to
a subsequent publication.

5 Perspectives

In several parts of physics and in particular in mechanics there are many models
which do not contain any adjustable parameter. For instance if one neglects external

perturbations the vibrations of a pendulum are determined by its length, its mass, and the acceleration of gravity; the important point being that all these parameters can be measured in an independent way through separate experiments. Most econometric models in contrast contain a huge number of adjustable parameters. One of the main innovations of econophysics was precisely to try to keep the number of adjustable parameters at a minimum. Nevertheless even the simplest stock market models (e.g. simple forms of minority games) contain some parameters that cannot be determined by separate observations. As a result it is difficult to carry out significant tests.

In proposing a dynamical framework for price peaks we had two main objectives: (i) to propose some elements for a unified theory of speculation, that is to say a theory which is not confined to financial markets; (ii) to emphasize the role played by the response times of economic agents and analyze the implications of their different orders of magnitude.

Some of our tests were merely qualitative, others were more detailed. What is most needed now is additional *comparative* evidence which will permit us to determine response times and other dynamical parameters with more accuracy and which will progressively enable us to sharpen the picture.

In the above program the word "comparative" is of cardinal importance. Huge amounts of statistical data are available in statistical yearbooks or on financial websites but, very little of this evidence can be used in a comparative way.

Because of the requirements of professional specialization, adopting a comparative perspective is not an obvious step: for instance the bond market specialist will not know what happens in real estate markets and the property developer will have only a faint notion of what happens on stock markets. As a matter of fact this obstacle is not proper to economics or finance. Thus, the compartimentalization of historical studies into various subfields has long prevented the development of truly comparative studies; this is unfortunate for, as we have tried to show elsewhere (Roehner and Syme 2002), a comparative perspective may throw new light on many seemingly unrelated phenomena. In a sense we pursued the same objective in the present book: by drawing a parallel between similar speculative phenomena occurring in different economic spheres and in various time periods we tried to come closer to a unified view of speculation. I am convinced that, if carried on consistently and systematically, such an approach will be really fruitful and will permit more and more accurate statistical tests.

Main data sources

Throughout this book we emphasized the crucial importance of an easy access to various data sources; they play the same role in econophysics as experimental data handbooks in physics. This section is devoted to a brief description of the main sources used in this book. But first of all it must be emphasized that the Internet opened a new era for comparative social research; it brought about a revolution of the same magnitude as the invention of the microscope in biology. Indeed the Internet gives free access to sources which provide a level of detail which would have been inconceivable in printed publications. As an example one can mention the fact that the CDC (Center of Disease Control) Wonder data base provides data for the death by suicide (or other causes of death) for all the counties in the United States (numbering over 3,100), by age and by year – in printed form such a data set would represent tens of thousands of pages which is why it has never been published in such detail.

In coming years such data bases will become more common in a increasing number of countries, thus providing an unique opportunity for comparative studies. In this sense it can be said that the Internet will broaden the horizon of social research in the same way as the Hubble Space Telescope broadened the horizon of astronomers and astrophysicists in the 1990s.

While some of the sources (such as national statistical yearbooks) that we used can be found in any major library, others, such as the sources for stamp or book prices, are not so easy to locate. For obvious reasons it was not possible to reproduce these data in the present volume; however the information given below should help readers to find the data they are looking for. In this section we only list the "main" sources corresponding to major data sets; other separate statistical figures have occasionally been used which came from sources listed in the general reference section. For each item we have given the full title, a brief description of the content and the place where it can be found. The following abbreviations have been used for that purpose.

[All] Can be found in all major libraries
[BNF] French National Library (Paris 75013)
[Cujas] University Library for Economics and Law (Cujas street, Paris 75005)
[FNSP] Library of the National Foundation for Political Science (Paris 75006)
[INSEE] Library of the Institut National de la Statistique et des Etudes Economiques (Paris – Malakoff)
[LC] Library of Congress (Washington)

American Book-Prices Current (since 1894). New York: E.P. Dutton (and other publishers subsequently).
  Content: Description and price of antiquarian books sold at auctions in the United States. [BNF]
Annuaire Statistique de la France. Published by the Institut National de la Statistique et des Etudes Economiques.
  Content: National statistical yearbook of France. [All]
Beacon Management: Website. Address: http://www.beaconmgmt.com
  Content: Data about the venture capital industry.
Book-auction records (formerly known as "Sales records"), a priced and annotated annual record of London book auctions. Edited in 1902 by F. Karslake and by other editors subsequently. The subtitle has somewhat evolved in the course of time becoming: a priced and annotated record of international book auctions.
  Content: Description and price of antiquarian books sold at auctions. [BNF]
Bulletin de la Statistique Générale de la France et du Service d'Observation des Prix. It is the earlier title of the above publication. [INSEE]
Bulletin Mensuel de Statistique. Published by the Institut National de la Statistique et des Etudes Economiques.
  Content: Monthly and quarterly French economic statistics. [INSEE]
Chambre des Notaires de Paris. Since 1985 the Chambre des Notaires (lawyer syndicate) de Paris has collected and published data on the price of apartments in Paris. They became available on the Internet in 2000 at the address: http://www.paris.notaires.fr
Commercial and Financial Chronicle. Weekly publication starting in 1896.
  Content: Gives weekly prices for all American stocks as well as other data about financial markets. [Cujas]
Common Stock Price Histories 1910–1986. WIT Financial Publishers (1987). Logarithmic supplement (1988).
  Content: Gives graphs of stock prices of almost all American companies. [LC]
Documents sur le problème du logement à Paris (1946). Ministère de l'économie. Imprimerie Nationale. Paris.
  Content: Data on the price of houses in Paris. [BNF]
FDA Consumer. The official magazine of the Food and Drug Administration. Available on Lexis-Nexis, an electronic newspaper data base.
Historical Statistics of the United States published by the US Department of Commerce (1975).
  Content: Historical continuation of statistical series published in the *Statistical Abstract of the United States.* [All]
Information Access Company. Periodical available on Lexis-Nexis, an electronic newspaper data base.
International Financial Statistics. Published by the International Monetary Fund.
  Content: Economic series for many countries. [All]

Jahrbuch der Bücherpreise. Edited by Otto Harrassowitz.
   Content: Description and price of antiquarian books sold at auctions in Germany.
   [BNF]
Japan Statistical Yearbook
   Content: National statistical yearbook for Japan. [All]
Jetro Business Facts and Figures. Jetro means: Japan External Trade Organization.
   Content: Economic statistics [Available at Jetro-Paris: 151 bis rue St Honoré]
Le Marché Immobilier Français: Les chiffres et les sources. (French periodical).
   Content: Data about the French real estate market. [BNF]
Main Economic Indicators. Historical Statistics 1969–1988. OECD.
   Content: Statistical series for the OECD countries. [All]
Monthly Commodity Price Bulletin 1960–1984 and 1970–1989 published by the United
   Nations Conference on Trade and Development (UNCTAD)
   Content: Monthly prices of commodities. [All]
Moody's Investors Service. Website: http://Moodys.com
Mutual Fund Factbook. Available on the website: http://ici.org
Ramses 1981: Rapport annuel mondial sur le système économique et les stratégies.
   Economica. Paris.
   Content: Comparative economic data. [FNSP]
Statistical Abstract of the United States. Published by the U.S. Department of Commerce.
   Content: National statistical yearbook for the United States. [All]
Statistisches Jahrbuch für das deutsche Reich.
   Content: National statistical yearbook for Germany (period before 1946). [Cujas]
Survey of Current Business.
   Content: Quarterly American statistical data. [All]
The World Almanac and Book of Facts (published since 1923). Press Pub. Co. New York.
   Content: Various statistical data (such as for instance the number of cars or boiler
   explosions). [FNSP]
Towsend Letters for Doctors and Patients. Periodical available on Lexis-Nexis, an
   electronic newspaper data base.
Venture Coach. Website: http://www.venture.coach.com
   Content: Data about the venture capital industry.
Yvert et Tellier stamp catalogs.
   Content: Price estimates for postage stamps (all countries). [BNF or Musée de la
   Poste, 34 Bld Vaugirard, Paris]

References

Abel, W. (1966). *Crises agraires en Europe (XIIIe–XXe siècle)*. Paris: Flammarion. English translation (from the German) (1980). *Agricultural Fluctuations in Europe from the Thirteenth to the Twentieth Centuries*. London: Methuen.

Allais, M. (1955). Fondements d'une théorie positive des choix comportant un risque et critique des postulats de l'école américaine. *Annales des Mines* (special issue).

Allais, M. (1968). L'économique en tant que science [A scientific view of economics]. *Revue d'Economie Politique*.

Allais, M. (1997). *L'anisotropie de l'espace: la nécessaire révision de certains postulats de la théorie contemporaine. 1. Les données de l'expérience.* Paris: Juglar.

Allais, M. and Hagen, O. (eds) (1979). *Expected Utility Hypothesis and the Allais Paradox.* Dordrecht: Reidel Publishing.

Allison, J.W. (1983). *Behavioral Economics.* New York: Praeger.

Andrew, A.P. (1907). The influence of crops upon business in America. *Quarterly Journal of Economics*, **20**, 323–353.

Ansidei, M., Carassus, J., and Strobel, P. (1978). *Logements: pourquoi la hausse des prix. Evolution 1960–1976*, La Documentation Française.

Atkin, M. (1989). *Agricultural Commodity Markets: A guide for Futures Trading.* London: Routledge.

Bachelier, L.J.B.A. (1900). Théorie de la spéculation. *Annales de l'Ecole Normale Supérieure*, 3, Paris: Gauthier-Villars. English translation in Cootner, P. (ed.) (1964). The Random Character of Stock Market Prices.

Baulant, M. and Meuvret, J. (1960). *Prix des céréales extraits de la mercuriale de Paris (1520–1698)*. Paris: S.E.V.P.E.N.

Bendix, R. (1978). *Kings or People: Power and the Mandate to Rule*. Berkeley, CA: University of California Press.

Bernanke, B.S. (1983). Non-monetary effects of the financial crisis in the propagation of the Great Depression. *American Economic Review*, **73**, 257–276.

Biollay, L. (1885). *Le pacte de famine*, Paris: Guillaumin.

Blume, M.E., Siegel, J.J., and Rottenberg, D. (1993). *Revolution on Wall Street: The Rise and Decline of the New York Stock Exchange*. New York: W.W. Norton.

Bouchaud, J.-P. and Mézard, M. (2000a). Wealth condensation in a simple model of economy. Preprint Cond-mat/0002374.

Bouchaud, J.-P. and Potters, M. (2000b). *Theory of Financial Risk*. Cambridge: Cambridge University Press.

Bouchaud, J.-P., Marsili, M., Roehner, B.M., and Slanina, F. (eds) (2001). Proceedings of the Prague 2001 Conference on Econophysics (to appear).

Bourgeois, R. (1913). *La crise égyptienne*. Paris: Arthur Rousseau.

Bouvier, J. (1960). *Le krach de l'Union Générale (1878–1885)*. Paris: Presses Universitaires de France.

Boyajian, W.E. (1988a). An economic view of the past decade in diamonds. *Gems and Gemology*, **24**, 134–147.

Boyajian, W.E. (1988b). World diamond production. *Gems and Gemology*, **24**, 148–153.

Bruchey, S. (1991). *Modernization of the American Stock Exchange 1971–1989*. New York: Garland Publishing.

Byron, F.W. and Fuller, R.W. (1970). *Mathematics of Classical and Quantum Physics*, Vol. 2. Reading, MA: Addison-Wesley.

Calomiris, C.W. and Gorton, G. (1991). The origins of banking panics: models, facts, and bank regulation. In *Financial Markets and Financial Crises*, ed. R.G. Hubbard. Chicago, IL: NBER. University of Chicago Press.

Carson, C. (ed.) (1999). *The Autobiography of Martin Luther King, Jr*. London: Little, Brown & Co.

Chalmin, P. (ed.) (1999). *Cyclope: matières premières et commodités*. Paris: Economica.

Chapman, S.J. and Knoop, D. (1906). Dealing in futures on the cotton market. *Journal of the Royal Statistical Society*, **69**, 2, 321–373.

Chaussinand-Nogaret, G. (1970). *Les financiers du Languedoc au 18e siècle*. Paris: S.E.V.P.E.N.

Chiang, A.C. (1984). *Fundamental Methods of Mathematical Economics*. Tokyo: McGraw-Hill.

Cole, A.H. (1927). Cyclical and sectional variations in the sale of public lands 1816–1860. *The Review of Economic Statistics*, **9**, 41–53.

Cole, A.H. and Frickey, E. (1928). The course of stock prices, 1825–1866. *The Review of Economic Statistics*, **10**, 117–139.

Common Stock Price Histories 1910–1986. Morristown, NJ: WIT Financial Publishers (1987), Logarithmic Supplement (1988).

Cont, R. and Bouchaud, J.-P. (2000). Herd behavior and aggregate fluctuations in financial markets. *Macroeconomic Dynamics*, **4**, 170.

Davies, J.B. and Shorrocks, A.F. (2000). The distribution of wealth. In *Handbook of Income Distribution*, ed. A.B. Atkinson and F. Bourguignon. Amsterdam: Elsevier.

Deaton, A. and Laroque, G. (1992). On the behavior of commodity prices. *Review of Economic Studies*, **59**, 1–23.

Descartes, R. (1637 [1824]). *Oeuvres de Descartes*, Vol. 5, ed. Victor Cousin. Paris: F.G. Levrault. English translation (1965). *Discourse on Method, Optics, Geometry and Meteorology*. Boobs-Merill Co.

Doerflinger, T.M. and Rivkin, J.L. (1987). *Risk and Reward: Venture Capital and the Making of America's Great Industries*. New York: Random House.

Doyle, A.C. (1910). *A Scandal in Bohemia*. Chicago.

Drame, S., Gonfalone, C., Miller, J.A., and Roehner, B. (1991). *Un siècle de commerce du blé en France 1825–1913: les fluctuations du champ des prix*. Paris: Economica.

Duthy, R. (1978). *Alternative Investment: A Guide to Opportunity in the Collectibles Market*. New York: Times Books.

Emery, H.C. (1896). *Speculation on the Stock and Produce Exchanges of the United States*. New York: University of Columbia Press.

Encyclopédie Internationale des Sciences et des Techniques (1973). Paris: Les Presses de la Cité.

Fama, E.F. (1965). The behavior of stock-market prices. *Journal of Business*, **38**, 34–105.

Farmer, J.D. (2002). Market force, ecology, and evolution. *Journal of Economic Behavior and Organization* (to appear).

Farrel, M.L. (ed.) (1972). *The Dow Jones Averages 1885–1970*. Princeton, NJ: Dow Jones.

Fay, S. (1982). *The Great Silver Bubble*. London: Hodder and Stoughton.

Fedenia, M. and Grammatikos, T. (1992). Options trading and bid-ask spread of the underlying stocks. *Journal of Business*, **65**, 335–351.

Feuilloley, M. (1996). Spécificités des objets d'art et de collection comme actifs patrimoniaux. Ph.D Thesis. University of Lille.

Friedman, M. (1953). *Essays in Positive Economics*. Chicago, IL: The University of Chicago Press.

Galilei [Galileo] (G.) (1730). *Mathematical discourses concerning two new sciences relating to mechanics and local motion in four dialogues. I: Of the resistance of solids against fraction. II: Of the cause of their coherence. III: Of local motion, viz. equable and naturally accelerate. IV: Of violent motion or of projects. With an appendix concerning the center of gravity of solid bodies*. London: J. Hooke.

Geoffron, P. (1990). Le processus de formation d'une innovation financière: le capital-risque. Une analyse comparative des modèles américains et français. Thesis. University Paris 13. Paris-Villetaneuse.

Gneezy, U. (1997). *Essays in Behavioral Economics*. Center for Economic Research. Tilburg University.

Gofman, J.W. (1985). *X-rays: Health Effects of Common Exams*. Sierra Club Books.

Gopikrishnan, P., Meyer, M., Amaral, L.A.N., and Stanley, H.E. (1998). Inverse cubic law for the distribution of stock price variations. *European Physical Journal B*, **3**, 139–140.

Gopikrishnan, P., Rosenow, B., Plerou, V., and Stanley, H.E. (2000). Identifying business sectors from stock price fluctuations. Preprint. Cond-mat/0011145.

Gottschalk, P. and Smeeding, T.M. (2000). Empirical evidence on income inequality in industrialized countries. In *Handbook of Income Distribution*, ed. A.B. Atkinson and F. Bourguignon. Amsterdam: Elsevier.

Granger, C.W.J. (1991). Reducing self-interest and improving the relevance of economic research. Paper presented at the 9th International Congress of Logic, Methodology, and Philosophy of Science. Uppsala (Sweden).

Grimmet, G.R. and Stirzaker, D.R. (1982). *Probability and Random Processes*. Oxford: Clarendon Press.

Hautcoeur, P.-C. (1997). Le marché financier français entre 1870 et 1900. In: *La longue stagnation en France* 1873–1897, ed. Breton (Y.) *et al*. Paris: Econometrica.

Johansen, A. and Sornette, D. (1999a). Financial "Anti-bubbles": log-periodicity in gold and Nikkei collapses. *International Journal of Modern Physics C*, **10**, 4, 563–575.

Johansen, A. and Sornette, D. (1999b). Log-periodic power law bubbles in Latin-American and Asian markets and correlated anti-bubbles in Western stock markets: An empirical study. Cond-mat/9907270.

Jones, C.M. (2000). A century of stock market liquidity and trading costs. Preprint (November) Columbia University.

Juglar, C. (1862). *Des crises commerciales et de leur retour périodique en France, en Angleterre et aux Etats-Unis*. English adaptation (1893, 1966). *A Brief History of Panics and Their Periodical Occurrence in the United States*. New York: A.M. Kelley.

Karpoff, J.M. (1987). The relation between price changes and trading volume: a survey. *Journal of Financial and Quantitative Analysis*, **22**, 1, 109–126.

Katona, G. (1980). *Essays in Behavioral Economics*. Institute for Social Research. University of Michigan.

Kelly, M. and O'Grada, C. (1999). Market contagion. Evidence from the panics of 1854 and 1857. Workpaper. Centre for Economic Research Dublin. *American Economic Review*, **90**, 5, 1110–1124.

Labrousse, E., Romano, R., and Dreyfus, F.-G. 1970: *Le prix du froment en France au temps de la monnaie stable* (1726–1913). Paris: S.E.V.P.E.N.

Langley, J. and Langley, S. (1989). State-level wheat statistics, 1949–1988. *Statistical Bulletin Number 779*. United States Department of Agriculture.

Leontief, W. (1982). Academic economics. *Science*, **9**, 17, 104–107.

Leontief, W. (1993). Can economics be reconstructed as an empirical science? *American Journal of Agricultural Economics*, October, 2–5.

Lieberson, S. (2000). *A Matter of Taste. How Names, Fashions, and Culture Change*. New Haven Yale University Press.

Liesner, T. (1989). *One Hundred Years of Economic Statistics*. New York: Facts on File.

Lillo, F. and Mantegna, R.N. (2000). Statistical properties of statistical ensembles of stock returns. *International Journal of Theoretical and Applied Finance*, **3**, 3, 405–408.

Lillo, F. and Mantegna, R.N. (2001). Ensemble properties of securities traded on the NASDAQ market. Proceedings of the Prague 2001 Conference on Econophysics (to appear).

Lucier, G., Budge, A., Plummer, C., and Spurgeon, C. (1991). US potato statistics 1949–1989. Statistical Bulletin Number 779. United States Department of Agriculture.

Lux, T. and Marchese, M. (1999). Scaling and criticality in a stochastic multi-agent model of a financial market. *Nature*, **397**, 498–500.

Macaulay, F.R. (1938). Some theoretical problems suggested by the movements of interest rates, bond yields and stock prices in the United States since 1856. National Bureau of Economic Research. New York.

Majorana, E. (1942). Il valore delle leggi statistiche nelle fisica e nelle scienze sociali [On the importance of statistical laws in physics and in the social sciences]. *Scientia*, **36**, 58–66

Mandelbrot, B.B. (1963). The variation of certain speculative prices. *Journal of Business*, **36**, 394–419

Mandelbrot, B.B. (1975). *Les objets fractals: forme, hasard et dimension*. Paris: Flammarion. English translation (1977). *Fractals: Form, Chance and Dimension*. San Francisco: Freeman.

Mantegna, R.N. (1991). Levy walks and enhanced diffusion in Milan Stock-Exchange. *Physica A*, **179**, 232–242.

Mantegna, R.N. and Stanley, H.E. (2000). *An Introduction to Econophysics. Correlations and Complexity in Finance*. Cambridge: Cambridge University Press.

Marcis, R., West, S., and Leonard-Chambers, V. (1995). Mutual fund shareholders response to market disruptions. *Perspective*, **1**, 1 (July), published by the Investment Company Institute.

Martin, G. (1908). Les famines de 1693 et 1709 et la spéculation sur les blés. *Bulletin du Comité des Travaux Historiques et Scientifiques*. Section des Sciences Economiques et Sociales, 150–172.

Maslov, S. and Mills, M. (2001). Price fluctuations from the order book perpective: empirical facts and a simple model. Proceedings of the Prague 2001 Conference. *Econophysics* (to appear).

Massacrier, A. (1978). *Prix des timbres-poste français classiques de 1904 à 1975*. Paris: A. Maury.

McConnel, S. (1957). *And Deliver us from Inflation*. Toronto: Burns and MacEachern.

Meeker, J.E. (1922). *The Work of the Stock Exchange*. New York: Ronald Press Company.

Meuvret, J. (1971). Etudes d'histoire économique. *Cahiers des Annales*, No 32.

Miller, J.A. (1999). *Mastering the Market. The State and the Grain Trade in Northern France*. Cambridge: Cambridge University Press.

Mishkin, F.S. (1991). Asymmetric information and financial crises: a historical perspective. In *Financial Markets and Financial Crises*, ed. R.G. Hubbard. Chicago: University of Chicago Press.

Mitchell, B.R. (1988). *British Historical Statistics*. Cambridge: Cambridge University Press.

Modeste, V. (1862). *De la cherté des grains et des préjugés populaires qui déterminent les violences dans les temps de disette*. Paris: Guillaumin.

Montroll, E.W. and Badger W.W. (1974). *Introduction to Quantitative Aspects of Social Phenomena*. New York: Gordon and Breach.

Morgan, E.V. and Thomas, W.A. (1962). *The Stock Exchange, Its History and Functions*. London: Elek Books.

Mouton, C. and Chalmin, P. (1985). *Matières premières et échanges internationaux*, Vol. 5. Paris: Economica.

Nerlove, M. (1958). Adaptative expectations and cobweb phenomena. *Quarterly Journal of Economics*, **72**, 227–240.

Newcomb, S. (1886). *Principles of Political Economy*. New York: Harper.

North, D. (1961). *Economic Growth in the United States*. Englewood Cliffs, NJ: Prentice Hall.

North, D. (1966). *Growth and Welfare in the American Past: A New Economic History*. Englewood Cliffs, NJ: Prentice Hall.

Onsager, L. (1944). A 2d model with an order-disorder transition. *Physical Review*, **65**, 117–149.

Papoulis, A. (1965). *Probability, Random Variables, and Stochastic Processes*. Tokyo: McGraw-Hill.

Pareto, V. (1917). *Traité de sociologie générale*. Lausanne: Payot. English translations (1935). *The Mind and Society*. 4 Vols. New York: Harcourt, Brace. (1963). *The Mind and Society: A Treatise on General Sociology*. New York: Dover.

Parzen, E. (1960). *Modern Probability Theory and Its Applications*. Tokyo: Wiley.

Plerou, V., Amaral, L.A.N., Gopikrishnan, P., Meyer, M., and Stanley, H.E. (1999). Similarities between the growth dynamics of university research and of competitive economic activities. *Nature*, **400**, 6743, 433–437.

Plerou, V., Gopikrishnan, P., Amaral, L.A.N., Meyer, M., and Stanley, H.E. (2001). Scaling of the distribution of price fluctuations of individual companies. Cond-mat/9907161 (revised February 1, 2001).

Powers, M.J. (1970). Does futures trading reduce price fluctuations in cash markets? *American Economic Review*, **60**, 3, 460–464.

Priestley, M.B. (1981). *Spectral Analysis and Time Series. Vol.1. Univariate series*. New York: Academic Press.

Putnam, R.D. (2000). *Bowling Alone. The Collapse and Revival of American Community*. New York: Simon & Schuster.

Quételet, A. (1835). *Sur l'homme et le développement de ses facultés: ou essai de physique sociale*. 2 Vols. Bachelier. Paris. English translation (1968). *A Treatise on Man and the Development of His Faculties*. New York: B. Franklin.

Quételet, A. (1841). Résumé des observations sur la météorologie, sur le magnétisme, sur les températures de la terre, sur la floraison des plantes, faites à l'Observatoire Royal de Bruxelles en 1840 par le directeur A. Quételet. Mémoires de l'Académie Royale de Bruxelles. Vol.14.

Quételet, A. (1869). *Physique sociale: ou essai sur le développement des facultés de l'homme.* 2 Vols. Brussels: C. Muquardt.

Quételet, A. (1870). Détermination de la déclinaison et de l'inclinaison magnétique et l'occultation de Saturne par la Lune le 19 avril 1870. *Bulletin de l'Académie Royale de Belgique*, **29**, 450–455.

Quid 1997: [An annual compilation of data] edited by D. and M. Frémy. Robert Laffont. Paris.

Rea, J. and Marcis, R. (1996). Mutual fund shareholder activity during US stock market cycles 1944–1995. *Perspective*, **2**, 2 (March), published by the Investment Company Institute.

Reid, B. (2000). The 1990s: a decade of expansion and change in the US mutual fund industry. Perspective (July). Published by the Investment Company Institute.

Reznikov, S. (1990). Les envolées de la Bourse de Paris au XIXe siècle. *Etudes et Documents*, **2**, 223–244.

Richmond, P. (2000). Chance to dream. *Europhysics News*. September–October.

Roehner, B. (1982a). Order transmission efficiency in large hierarchical organizations. *International Journal of Systems Science*, **13**, 5, 531–46.

Roehner, B. (1984). Macroeconomic regularities in the growth of nations: an empirical inquiry. *International Journal of Systems Science*, **15**, 9, 917–936.

Roehner, B.M. (1989). Peaks and troughs in wheat price series. Working paper PAR LPTHE 89.

Roehner, B.M. (1995). *Theory of Markets: Trade and Space-Time Pattern of Price Fluctuations: A Study in Analytical Economics.* Berlin: Springer-Verlag.

Roehner, B.M. (1996). The role of transportation costs in the economics of commodity markets. *American Journal of Agricultural Economics*, **78**, 339–353.

Roehner, B.M. (1997). The comparative way in economics: a reappraisal. *Economie Appliquée*, **50**, 4, 7–32.

Roehner, B.M. (1999a). Spatial analysis of real estate price bubbles: Paris 1984–1993. *Regional Science and Urban Economics*, **29**, 73–88.

Roehner, B.M. (1999b). The space-time pattern of price waves. *The European Physical Journal B*, **8**, 151–159.

Roehner, B.M. (2000a). Determining bottom price levels after a speculative peak. *The European Physical Journal B*, **17**, 341–345.

Roehner, B.M. (2000b). The correlation length of commodity markets. 1. Empirical evidence 2. Theoretical framework. *The European Physical Journal B*, **13**, 175–187 and 189–200.

Roehner, B.M. (2001a). *Hidden Collective Factors in Speculative Trading: A Study in Analytical Economics.* Berlin: Springer-Verlag.

Roehner, B.M. (2001b). To sell or not to sell? Behavior of shareholders during price collapses. To appear in the *International Journal of Modern Physics C*.

Roehner, B.M. and Shiue, C.H. (2001). Comparing the correlation length of grain markets in China and France. *International Journal of Modern Physics C*, **11**, 7, 1383–1410.

Roehner, B.M. and Sornette, D. (1998). The sharp peak – flat trough pattern and critical speculation. *The European Physical Journal B*, **4**, 387–399.

Roehner, B.M., Sornette D. (2000): "Thermometers" of speculative frenzy. *The European Physical Journal B*, **16**, 729–739.

Roehner, B.M. and Syme, T. (2002). *Pattern and Repertoire: An Introduction to Analytical History*. Cambridge, MA: Harvard University Press (to appear).

Roehner, B. and Wiese, K.E. (1982). A dynamic generalization of Zipf's rank-size rule. *Environment and Planning A*, **14**, 1449–1467.

Sahlström, P. (2000). The effects of stock derivative markets on the underlying stock market: a study of the option markets using Finnish markets as a laboratory. *Acta Wasaensia*, No. 79.

Samuelson, P.A. (1952). Spatial price equilibriium and linear programming. *American Economic Review*, **42**, 283–303.

Schumpeter, J. (1933). The common sense in econometrics. *Econometrica*, **1**, 5–12.

Schwartz, A.J. (1995). An interview with Anna J. Schwartz. *The Newsletter of the Cliometric Society*, **10**, 2, 3–7.

Scott, A. (1977). *Neurophysics*. New York: Wiley.

Seuffert, G.K.L. (1857). *Statistik des Getreide und Viktualien Handels im Königreiche Bayern*. Munich: Weisz.

Shiller, R.J. (2000). *Irrational exuberance*. Princeton, NJ: Princeton University Press.

Simon, H.A. (1959). Theories of decision-making in economics and behavioral science. *American Economic Review*, **49**, 3, 253–283.

Simon, H.A. (1962). The architecture of complexity. *Proceedings of the American Philosophical Society*, **106**, 6, 467–482.

Simon, H.A. (1982–1997). *Models of Bounded Rationality*. 3 Vols. Cambridge, MA: MIT Press.

Simon, H.A. (1983). *Reason in Human Affairs*. Stanford, CA: Stanford University Press.

Snooks, G.D. (1993). *Economics without Time: A Science Blind to the Forces of Historical Change*. London: Macmillan.

Sobel, R. (1987). *The New Game on Wall Street*. New York: John Wiley & Sons.

Solomon, S. and Richmond, P. (2001). Power laws of wealth, market order volumes and market returns. Cond-mat/0102423 (February).

Sornette, D. and Andersen, J.V. (2001). Quantifying herding during speculative financial bubbles. Cond-mat/0104341 (April).

Sornette, D., Johansen, A., and Bouchaud, J.-P. (1996). Stock market crashes, precursors and replicas. *Journal de Physique I France*, **6**, 167–175.

Souma, W. (2000). Universal structure of the personal income distribution. Preprint Cond.-mat./0011373 (November).

Stauffer, D. and Sornette, D. (1999). Self-organized percolation model for stock market fluctuations. *Physica A* 271, N3–4, 496–506.

Stukeley, W. (1752, 1936). *Memoirs of Sir Isaac Newton's Life*. London: Taylor & Francis.

Taniguchi, T. (1993). Japan's banks and the "bubble economy" of the late 1980s. Center for International Studies. Princeton University.

Tilly, C. (1993). *European Revolutions 1492–1992*. Oxford: Blackwell.

Tilly, L.A. (1983). Food entitlement, famine and conflict. *Journal of Interdisciplinary History*, **14**, 333–349.

Tilly, L.A. (1992). The decline and disappearance of the classical food riot in France. Working paper No. 147. Center for Studies of Social Change. New School of Social Research. New York.

Tinker, H. (1974). *A New System of Slavery: The Export of Indian Labour Overseas 1830–1920*. Oxford: Oxford University Press.

Tychonov, A.N. and Samarski, A.A. (1964). *Partial Differential Equations of Mathematical Physics*. Vol.1. San Francisco: Holden Day.

Usher, A.P. (1913). *The History of the Grain Trade in France*. Cambridge, MA: Harvard University Press.

Von Neumann, J. and Morgenstein, O. (1953) *Theory of Games and Economic Behavior.* Princeton: Princeton University Press.

Vuitry, A. (1885). Le désordre des finances et les excès de la spéculation à la fin du règne de Louis XIV et au commencement du règne de Louis XV. Paris: C. Lévy.

Walras, A. (1874). *Eléments d'économie politique pure; ou théorie de la richesse sociale*. Lausanne: Corbaz. English translation: *Elements of Pure Economics: Or the Theory of Social Wealth*. London: Allen & Unwin (1954).

Walvin, J. (1992). *Black Ivory*. London: Harper Collins.

Warren, G.F. and Pearson, F.A. (1937). The building cycle. *Fortune* (August)

Warren G.F. and Pearson, F.A. (1937). *World Prices and the Building Industry. Index Numbers of Prices of 40 Basic Commodities for 14 Countries and Material on the Building Industry*. New York: John Wiley & Sons.

Weber, N.A. (1972). Gardening ants: the Attines. The American Philosophical Society. Philadelphia.

Weidlich, W. and Haag, G. (1983). *Concepts and Models of a Quantitative Sociology: The Dynamics of Interacting Populations*. Berlin: Springer-Verlag.

White, E.N. (1990). When the ticker ran late: the stock market boom and crash of 1929. In *Crashes and Panics: The Lessons from History*, ed. E.N. White. Homewood, IL: Dow Jones-Irwin.

White, E.N. (ed.) (1996). *Stock Market: Crashes and Speculative Manias*. Cheltemham: Edward Elgar.

Wilson, S.D. (1992). *The Bankruptcy of America: How the Boom of the 80s became the Bust of the 90s*. Germantown, TN: Ridge Mills Press.

Wilson, J.W., Sylla, R.E., and Jones, C.P. (1990). Financial market panics and volatility in the long run 1830–1988. In *Crashes and Panics: The Lessons from History*, ed. E.N. White. Homewood, IL: Dow Jones-Irwin.

Wirth, M. (1883). *Geschichte der Handelskrisen*. Frankfurt am Main: Sauerländer's Verlag.

Wood, C. (1992). *The Bubble Economy. The Japanese Economic Collapse*. London: Sidwick & Jackson.

Index

Abel, W., 91
Allais, M., 4, 22, 28–29
Amaral, L., xvii, 198
Amazon.com, 69
American Bell, 87
American Ship, 201
antiquarian books, 51, 94, 97
Apple, 138
apples, 170
arbitrage, 211
Arthur, B., 67
ascending (versus descending) phase, 126,
   181
Atkin, M., 69
automobile industry, 84–85

Bachelier, L.J.B.A., 28
Badger, W.W., 28
banking system, 72–73
bank suspensions, 78
Bernard, C., xii, 12
bid–ask margin, 62, 142, 181
Biollay, L., 91
biophysics, 3, 25–26
biotechnology, 85, 102
block transaction, 61
blue chip, 90
Blume, M.E., 56
bond market, 125–126, 172–174
Boston group, 198
Bouchaud, J.-P., xvii, 35, 114
Bouvier, J., 139
Boyle, xi, xiii
brokers' forecasts, 123
Bruchey, S., 68

California, 116–117
call-option, 55, 66, 137
Calomiris, C.W., 73
capitalization (NYSE), 143
characteristic time, 175–176
Chaussinand-Nogaret, G., 55
Chicago Board Options Exchange, 67

Cisco Systems, 123
cluster of events, 6, 15, 83
cobalt, 19, 95
collectibles *see* antiquarian books, postage stamps
collection of events *see* cluster of events
collective behavior, 55
collective rationality, 48
commission rate, 57, 60, 63, 148–149
commodities *see* apples, cobalt, corn, cotton,
   diamonds, gold, palladium, platinum, potatoes,
   silver, soy beans, wheat
communication costs, 56
comparative analysis, 23
complexity, 10–14
concave function, 166
concentration, 55–56, 58
consumer confidence, 144–147
convex function, 166
Coriolis force, xii, 13, 23
corn, 170
corner battles, 65
Corning, 138
cotton, 170
coupon interest rate, 123
credit crunch, 93, 172

damped oscillations, 188–189
damping, 186–195
day trader, 142
Deaton, A., 49
De Beer's, 67
Debreu, G., 22
decimalization, 142
Dell Computer, 138
derivatives, 67
Descartes, R., 3, 8, 11
Deutsch, K., xiv, 48
diamonds, 17–19, 67, 169
diffusion process, 213
dinosaurs, 17
dispersion of peak times, 175
distribution of price changes, 183, 203
DJI *see* Dow Jones Industrials

227

Printed in the United States
By Bookmasters